Predictive Analytics with KNIME

Frank Acito

Predictive Analytics with KNIME

Analytics for Citizen Data Scientists

 Springer

Frank Acito
Indiana University
Bloomington, IN, USA

ISBN 978-3-031-45632-9   ISBN 978-3-031-45630-5   (eBook)
https://doi.org/10.1007/978-3-031-45630-5

This Springer imprint is published by the registered company Springer Nature Switzerland AG
The registered company address is: Gewerbestrasse 11, 6330 Cham, Switzerland

Paper in this product is recyclable.

I dedicate this work to my loving wife, Sandie, who has been patient, supportive, and understanding.

Preface

Over the past years, university programs in analytics have become very popular. One estimate is that nearly 250 graduate programs in analytics have been launched with more on the horizon. This book, *Predictive Analytics with KNIME*, is based on my over 40 years of teaching analytics to thousands of students at the Kelley School of Business at Indiana University. The book draws from my lectures, Excel demonstrations, handouts, and other materials I developed. Based upon my years of teaching this material, I have learned where students have difficulty and have developed strategies for helping them learn which I have incorporated into the book.

Several texts on analytics are available, but this text differs in several ways. One differentiating feature of my book is using KNIME as the principal analytics software. KNIME is an open-source, drag-and-drop package that is especially easy to use. Unlike R and Python, which are used in other textbooks, learning KNIME is much more intuitive and does not require coding. The drag-and-drop workflows created in KNIME are self-documenting, so users can see exactly which analysis steps were taken and in what order.

The graphical interface in KNIME is like those in well-known, but expensive, commercial programs such as SAS Enterprise Miner, IBM SPSS Modeler, and Microsoft's Azure Machine Learning. Despite being open-source, KNIME has thousands of easy-to-use nodes which can perform a vast number of analyses, including nodes for deep learning, text analytics, and ensemble models.

Since KNIME is open-source, readers can use it for their own work or in organizations. KNIME is unlike many other tools that require expensive annual licensing. Feedback from my students while teaching courses on analytics indicated that they are often disappointed that they could not use the software after the end of the course because licenses were too costly for them and even for many of the organizations where they worked.

Another advantage of KNIME is that it can incorporate R code. In a few instances, I use R for specialized tasks that are either unavailable in KNIME or enable more customization of the analyses. However, knowledge of R is not a requirement, and all the R codes will be provided.

While providing conceptual background on the algorithms, the text also includes intuitive explanations. For example, principal components analysis is usually presented as a series of matrix calculations using eigenvalue decomposition, and the mathematics do not make clear how this technique works. While I include the equations, I also demonstrate how an eigenvalue can be obtained in Excel by maximizing the variance of a weighted average of a set of variables. This is one example of how the book should be accessible to those with little technical background but still interesting to those more comfortable with statistics and mathematics.

Images of the KNIME workflows and detailed descriptions of the nodes in each KNIME workflow are provided. All the KNIME workflows discussed in the book, including the required data sets, are available to readers. Users can download the workflows, run them on their computers, experiment with modifications, and use these as templates for further work. The site to download the workflows is https:// tinyurl.com/KNIMEWorkflows.

Bloomington, IN, USA   Frank Acito

Contents

Chapter 1
Introduction to Analytics

Analytics involves the application of data-based models to enhance results, reduce costs, and reduce risk in both profit-making and non-profit organizations. Eric Siegel coined "The prediction effect" to describe how models make predictions about people, documents, actions, or entities at the individual level (Siegel, 2013). He further stated that applicable predictive models do not have to be perfectly accurate. All that is necessary is that predictions using analytics need to be better than the method used in the past.

Analytics has emerged as a catch-all term for various business intelligence and application-related initiatives. It sometimes refers to statistical and mathematical data analyses in specific content areas such as sales, services, supply chain and logistics, health care, and fraud detection. Across these content areas, analytics can also be classified into three basic types based on the end objective of an application: description, prediction, and prescription, in rough order of sophistication.

Descriptive analytics refers to examining and interpreting data to answer questions such as, "What happened?" or "What is currently happening?" Data visualizations such as bar charts, scatterplots, line graphs, box plots, and histograms are commonly used to derive insights from data sets. Furthermore, descriptive analytics utilize basic summary statistics such as measures of central tendency and spread as well as correlations to understand the data better.

Predictive analytics focuses on predicting continuous variables and classifying categorical outcomes. Before building predictive models, it is crucial to utilize descriptive analytics techniques to comprehend, clean, and prepare the data. Techniques such as multiple regression, decision trees, and neural networks are among the many predictive analytics models available.

Prescriptive analytics represents an advanced approach that provides guidance to decision-makers by answering the question, "What should be done?" Simulation, recommendation engines, and optimization methods are used to develop prescriptive models. However, accurate predictive models serve as essential prerequisites before creating prescriptive models.

F. Acito, *Predictive Analytics with KNIME*,
https://doi.org/10.1007/978-3-031-45630-5_1

While descriptive and prescriptive analytics will be featured in some examples, the principal focus of this book is on methods for predictive analytics. Predictive analytics can provide direct value to decision-makers. Descriptive analytics are essential, but while reports and dashboards based on descriptive analytics can provide understanding and insight, decision-makers are left to integrate the findings into their beliefs and experiences to make choices. Prescriptive analytics, on the other hand, can give managers guidance on actions to take.

1.1   The Growing Emphasis on Analytics

The 2021 Digital Readiness Survey found that 89% of IT respondents reported increased use of business analytics in the previous 2 years.[1] A 2022 survey of supply chain and manufacturing executives said that 22% had adopted predictive and prescriptive analytics, with more than 80% reported intentions to adopt it in the next 5 years.[2] A 2019 study found that 60% of healthcare executives used predictive analytics. The human resources function has also recently emerged as a leader in analytics applications (Davenport, 2019).

Three developments have created significant opportunities to use analytics to support and enhance decision-making in business, health care and medicine, government, and not-for-profit organizations. The three developments are:

- An explosion in the volume, variety, and velocity of data.
- Better, faster, and cheaper hardware and software.
- Growing demand for fact based, data supported decision making.

Volume, Variety, and Velocity of Data
More data is being collected, from more sources, than ever before. Data has not only increased in volume; it has also gained tremendous richness and diversity. For many years data analysis was mainly applied to structured quantitative data that was clearly formatted with defined fields from internally generated information about finance, production, employees, and sales. Today, richer analyses can lead to new insights by combining internal data with external data from governments, social media, mobile devices, and many other sources. Advances have made extracting unstructured data from text, images, video, and voice possible.

In 2001 analyst Doug Laney formulated the 3V framework to characterize the dimensions of the data explosion (Laney, 2001).[3] Laney noted that the amount of data is expanding in three ways:

---

[1] https://www.manageengine.com/the-digital-readiness-survey-2021

[2] "The 2022 MHI Annual Industry Report" 2022. (https://www.mhi.org/publications/report).

[3] The original 3V framework has been expanded by various authors to include terms such as "veracity"and"value".(https://www.techtarget.com/searchdatamanagement/definition/5-Vs-of-big-data).

- Volume: The amount or size of data. Big data is about volumes of data that are reaching unprecedented levels. Even the prefixes used to identify big data metrics have evolved to quantify the massive amounts of data measured in bytes: mega (10^6), giga (10^9), tera (10^{12}, peta (10^{15}), exa (10^{18}), zeta (10^{21}), and yotta (10^{24}).
- Variety: The diverse range of data, including structured databases, unstructured data not directly machine-readable, time series, spatial, natural language, media, and data from connected devices.
- Velocity: The speed at which data is received, stored, managed, and analyzed, in some cases, in real- time.

One data source that is growing in importance is embedded sensors that collect and transmit data from connected devices – such as industrial machines, automobiles, and home appliances. This phenomenon, called "The Internet of Things," has led to smarter home security systems, autonomous farm equipment, self-driving vehicles, and numerous other developments. Data is generated by everything from cameras and traffic sensors to heart rate monitors, enabling richer insights into human behavior. Retailers, for instance, are building complete customer profiles across all their touchpoints, while healthcare providers can now monitor patients remotely. In addition, companies in traditional industries such as natural resources and manufacturing are using sensors to monitor their physical assets and logistics. They can use that data to predict and prevent bottlenecks or maximize utilization.

Some of the data generated is a byproduct of using the internet, smartphones, and other connected devices. This has been labeled "data exhaust" for how it streams out, like the way car exhaust spews out behind motor vehicles.

There is no question that the explosion of data, especially personally identifiable data, must be carefully used. Governmental rules and regulations have not kept up with the technical capabilities, so it is incumbent on analysts to carefully balance applications with ethical considerations.

Better, Faster, Easier to Use, and Less Expensive Hardware and Software
About 70 years ago, the era of digital computers began. The computers of the fifties and sixties were huge, room-filling machines that were operated in climate-controlled environments. These machines required one or more operators to monitor performance continuously. Since then, exponential improvements in the basic circuitry, memory chips, and processors have produced computers that are smaller and cheaper than anything imagined in the 1960s.

The growth of computing power was noted in the 1960s by Gordon Moore, a Caltech professor and one of the founders of the giant Intel Corporation. Moore's Law (the observation that the number of transistors in an integrated circuit has doubled every 18–24 months since 1965) has tracked the exponential growth in the power of microprocessors since 1965. One interesting calculation compares one of the first all-transistor computers, the IBM 7090 produced in 1961, with today's powerful laptops. A modern chip in a high-end laptop can operate 30,000 times as fast as the IBM 7090 (which would cost $20 million in today's money versus

$500–$2000 for today's machines). Today's analysts have computing power on personal devices that can use the latest advanced techniques even with huge data sets.

Several developments have improved data mining and predictive analytics software tools, making them easier to use, more powerful, and, in some cases, less expensive. One key development has been the emergence of cloud computing, which provides software tools as web services and turns computational power into a utility. The cloud-based supply model enables on-demand power and pay-as-you-go. Leaders in this domain include Microsoft, Amazon Web Services, IBM, and Google.

Desktop software has also become more sophisticated, more powerful and user-friendly, lowering the entry barriers for building analytics models. Open-source tools such as Python and R provide free access to world-class tools for data analytics. Advanced data analysis packages used to have steep learning curves and, even when learned, required arcane commands and complex syntax. Today, analytics can be done with drag-and-drop tools that have made analytics accessible to more people. Furthermore, open-source tools such as KNIME, H2O, Orange, and Weka are free.

Many software programs can quickly produce basic visual representations of data such as bar charts, scatter plots, pie charts, and line graphs. Such visualizations are a tiny sliver of what is possible. Advanced data visualization is the last link in communicating results – from the computer to the human mind. Techniques are now used to show dynamic content and animations, allow real-time querying alerts, and combine multiple visualizations into interactive dashboards. Enhanced data visualization tools such as Power BI Desktop, Tableau, and Qlik enable the creation of state-of-the-art visualizations and dashboards for understanding and communicating insights from data.

Growing Demand for Fact-Based, Data-Supported Decision Making
Today, advanced analytics are used in all types of commercial and non-commercial organizations. However, while decision-makers know the benefits of data analytics, many organizations have yet to realize the potential advantages fully.

A very early example of predictive analytics was developed in 1689 by the Lloyd Company, which used past data to evaluate the risk of sea voyages and thus set premiums for insuring voyages.[4] A more recent example of commercial analytics applications was the system created by the Fair Isaacs Corporation in the 1950s with the development of FICO scores for credit scoring.[5] Since then, applications and opportunities have existed in all functional areas. For example, an emerging trend is using analytics in human resources, where applications are being developed to automatically filter job applications, deal with turnover issues, optimize and manage benefits, drive succession planning, and guide employee training.

---

[4] "A Brief History of Predictive Analytics." 2019. (https://medium.com/@predictivesuccess/a-brief-history-of-predictive-analytics-f05a9e55145f).

[5] "FICO History." (https://www.fico.com/en/history).

Greater management sophistication (and expectations) regarding the use of data to support decision-making has been a driver of analytics. Organizations are turning to analytics to make fact-based decisions, relying less on intuition and becoming more data driven. The influential McKinsey report, "Big data: The next frontier for innovation, competition, and productivity" (Manyika et al., 2016), provided an in-depth look at the landscape of big data and predicted the huge benefits that can be generated by mining the big data in domains such as health care, public service, government, retail, manufacturing, and logistics. The report also projected a shortage of expertise in data mining and data-sophisticated managerial talent.

1.2   Applications of Analytics

Analytics is being used in all sorts of creative and impactful ways. For example, Google launched Project Oxygen in 2008 to document the common behaviors of its managers. Analyses of more than 10,000 performance reviews were examined to determine the effects of leadership on employee retention. Applying insights from the investigations led to measurable improvements in turnover, satisfaction, and performance (Harrell & Barbato, 2018).

Analytics is used extensively in professional and collegiate football, basketball, soccer, and other sports to diagnose and improve individual and team performance, recruit new players, and increase revenue.

The lodging industry uses analytics for a wide variety of applications of analytics. Marriott, a leader in the use of analytics, uses data to identify new sources of revenue, cultivate new customers, increase the loyalty of legacy ones, and identify opportunities for new properties in new markets (Eisen, 2018).

Amazon has several recommendation engines for books, consumer products, music, and videos that predict the shopping preferences of their customers. While many retailers use recommendation tools, Amazon has one of the best. Rather than blindly suggesting a product, Amazon uses data analytics and machine learning to drive its recommendation engine. Amazon's system uses three types of data to decide which products to recommend to a consumer using its site: (1) customer demographics that are related to product preferences; (2) product-product relationships such as recommending books and movies based on the patterns of products purchased by others; (3) data on customers' prior purchases and search behavior. McKinsey estimated that in 2017 35% of Amazon's consumer purchases could be tied back to the company's recommendation system (MacKenzie et al., 2013).

Example Application to Health Insurance
An example of an analytics project by the consulting firm Accenture to improve operations in a health insurance provider was described in an article by Kumar et al. (2010). The project's focus was reducing errors in claims experienced by a health insurance company. These errors required extra administrative work, extra costs, and increased dissatisfaction by medical professionals and their patients due to

delays and incorrect denials of payments. Two common ways of identifying errors are a random selection of claims for audit and using rules devised by experts to flag claims with errors. While using expert-based rules is slightly better than random audits, the rules require investments in expert time for development and continued refinement. The Accenture project aimed to identify claims with errors (not necessarily fraud) before authorizing the final payment. To illustrate how predictive analytics was used to make the claims auditing process more efficient, the following example explains how analytics was used in this case. (The actual data used in the Accenture study is not available.)

Assume a hypothetical health insurance provider receives about 5000 claims every working day, resulting in 5000 claims × 250 days/year = 1,250,000 claims yearly. The company employs a team that takes a random sample of the audit claims. Experience indicates that about 5% of the claims audited have one or more errors, which means that about .05 × 1,250,000 = 62,500 claim errors occur each year for this provider. Also assume the audit team can review 200 randomly selected claims per day, so with 200 × 250 days, 50,000 claims can be audited each year. Using the assumed 5% error rate, the audit team would be expected to find about 2500 claim errors per year, only 4% of the 62,500 total. This meant 60,000 claim errors would not be flagged and require costly rework when a patient or provider disputes the claim and requests a second review.

A predictive model was developed using data from previous years. Each record in the database was a claim with patient and provider details, characteristics of the claim, and a flag indicating whether an error(s) was detected leading to rework was needed. This flag can be used as a binary target variable for a predictive model which uses variables from the claim text and data on the providers and patients. As with most models, the model cannot identify every error and may incorrectly flag claims without errors. Assume the model provides a score on each claim in the database reflecting the probability that the claim has an error and needs to be reviewed. The model's predictive accuracy was not perfect with many claims erroneously labeled as having errors and many that did have errors not correctly identified. Despite the errors, 25% of the claims with errors were correctly identified.

The model included a probability that a claim had an error and the data from the analysis was sorted from highest to lowest predicted probability of error. Assume that the top 50,000 claims with the highest chance of error are selected. (This matches the annual capacity, 50,000 audits per year, of the audit team.) The result is that in 25% × 50,000 = 12,500 claims with errors are presented to the audit team. This is five times the number of errors identified using the random sample.

1.3   The Citizen Data Scientist

Software tools that require little or no coding to develop sophisticated predictive analytics models make analytics available to a broader set of employees. Gartner analysts Tapadinhas and Idoine (2016) coined the term "citizen data scientist" to describe the emerging trend of democratizing analytics beyond dedicated, highly trained, and hard-to-find data scientists. Pidsley and Idoine (2021) defined a citizen data scientist as someone that uses analytics, but whose main job is not analytics.

Citizen data scientists might work in finance, accounting, marketing, operations, or human resources, but they could employ self-service analytics tools to increase the effectiveness and impact of their organizational work. Predictive analytics becomes an added activity in their everyday jobs.

Citizen data scientists can fill the gap left in IT organizations needing help to meet the demand for incorporating analytics into day-to-day decision-making and operations. Even deep learning tools, which have empowered computers to mimic human intelligence in highly specialized areas such as image analysis, speech processing, and text analytics, are available in accessible software such as KNIME.

The demand for highly skilled data scientists far outstrips the current supply. The following example illustrates this. Suppose an organization posted a job opening for a skilled data scientist. The ideal candidate would have 3 years of experience and skills in Python, SQL, Cloud Computing, and Machine Learning. According to Gartner Talent Neuron data, 98% of the current world talent pool would not qualify for the position, (Heizenberg, 2022). So, citizen data scientists are likely to have many opportunities to contribute while not replacing technical experts.

1.4   The Analytics Process

Developing a predictive analytics model is best approached as a deliberate sequence of activities that starts with carefully understanding the problem. These activities or steps are listed in sequence, but following the steps in a strict order rarely occurs in practice. Instead, iteration, revision, and repetition are usually required since the results of one step may require that previous actions be re-examined and redefined.

Several process frameworks have been published, including CRISP (Cross Industry Standard Process for Data Mining),[6] SEMMA (Sample, Explore, Modify, Model and Assess),[7] Knowledge Discovery in Databases (Fayyad et al., 1996), Microsoft's Team Data Science Process (TDSP, n.d.), and others. While the CRISP framework has not been updated since it was developed, and it has remained the most popular model based on Google search volume.[8]

---

[6]"IBM SPSS Modeler CRISP-DM Guide." (https://www.ibm.com/docs/it/SS3RA7_18.3.0/pdf/ModelerCRISPDM.pdf).

[7]"Data Mining and SEMMA." (https://documentation.sas.com/doc/en/emcs/14.3/n0pejm83csb-ja4n1xueveo2uoujy.htm#:~:text=SEMMA%20is%20an%20acronym%20used,tabs%20with%20the%20same%20names).

[8]"CRISP-DM is Still the Most Popular Framework for Executing Data Science Projects" https://www.datascience-pm.com/crisp-dm-still-most-popular/#:~:text=CRISP%2DDM%20comes%20out%20as%20the%20most%20popular&text=Note%20that%20CRISP%2DDM%20is,possible%20and%20indeed%2C%20often%20occur

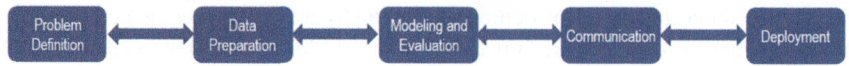

Fig. 1.1 A revised analytics framework

Based on the author's experience in developing predictive models, some modifications were made to the standard CRISP model for purposes of this book, resulting in the revised framework shown in Fig. 1.1. The changes are:

- The label on the first step is changed to Problem Definition, since this more clearly indicates that the output of this step needs to state the problem to be solved by the analysis task. Business understanding is implied in this step.
- The two steps, Data Understanding and Data Preparation, are combined into a single step Data Preparation, since the two separate steps shown in the CRISP model are rarely conducted in the sequence shown but rather tend to be done jointly.
- Modeling and Evaluation are combined into a single step to reflect the iteration typically needed in building an accurate model.
- A new step, Communication, has been added to emphasize the importance of gaining understanding and buy-in from others, which is usually needed if the model is to be deployed and used.
- Deployment is the final step in the revised model as is true of the CRISP model.

The modified CRISP model from Fig. 1.1 was used to organize the chapters covered in the book, starting with Problem definition in Chap. 2 and ending with Communication and deployment in Chap. 14 (with a detour in Chap. 3 to introduce KNIME).

1.5 Summary

Analytics involves using data models to enhance results, reduce costs, and reduce risk. Its use has grown rapidly in recent years driven by three key developments: the explosion of data volume, variety, and velocity; advances in hardware and software; and a growing demand for data-driven decision making. Applications of analytics span many industries including tech, retail, healthcare, sports, and more. Advances have enabled "citizen data scientists" without extensive statistical training to develop predictive models using self-service tools which help meet the high demand for analytics expertise. Developing analytics models involves a series of steps beginning with problem definition and ending with deployment.

References

Davenport, T. (2019). *Is HR the most analytics-driven function?* https://hbr.org/2019/04/is-hr-the-most-analytics-driven-function. Accessed 29 July 2023.

Eisen, D. (2018). *Marriott bets on predictive analytics for brand growth.* https://www.hotelmanagement.net/tech/marriott-builds-its-brands-by-knowing-more-about-you. Accessed 29 July 2023.

Fayyad, U., Piatetshy-Shapiro, G., & Smyth, P. (1996). *The KDD process for extracting useful knowledge from volumes of data mining communications of the ACM.* https://dl.acm.org/doi/pdf/10.1145/240455.240464. Accessed 29 July 2023.

Harrell, M., & Barbato, L. (2018). *Great managers still matter: The evolution of Google's Project Oxygen.* https://rework.withgoogle.com/blog/the-evolution-of-project-oxygen/. Accessed 29 July 2023.

Heizenberg, J. (2022). *How to attract and retain data & analytics talent.* https://www.itworldcanada.com/blog/how-to-attract-and-retain-data-analytics-talent/470353. Accessed 29 July 2023.

Kumar, R. Ghani, M., & Mei, Z. (2010). Data mining to predict and prevent errors in health insurance claims processing. In *Proceedings of the 16th ACM SIGKDD international conference on knowledge discovery and data mining* (pp. 65–74).

Laney, D. (2001). *3d data management: Controlling data volume, velocity and variety.* Institute of TechnologMETA Group Research Note. https://studylib.net/doc/8647594/3d-data-management%2D%2Dcontrolling-data-volume%2D%2Dvelocity%2D%2Dan... Accessed 29 July 2023

MacKenzie, I., Meyer, C., & Nobel, S. (2013). *How retailers can keep up with consumers.* https://www.mckinsey.com/industries/retail/our-insights/how-retailers-can-keep-up-with-consumers. Accessed 29 July, 2023.

Manyika, J., Chui, M., Brown, B., Bughin, J., Dobbs, R., Roxburgh, C., & Byers, A. H. (2016). *Big data: The next frontier for innovation, competition, and productivity.* McKinsey Digital. https://www.mckinsey.com/capabilities/mckinsey-digital/our-insights/big-data-the-next-frontier-for-innovation. Accessed 29 July 2023.

Pidsley, D., & Idoine, C. (2021). *Maximize the value of your data science efforts by empowering citizen data scientists.* https://www.gartner.com/en/documents/3878963. Accessed 29 July 2023.

Siegel, E. (2013). *Predictive analytics: The power to predict who will click, buy, lie, or die.* Wiley.

Tapadinhas, J., & Idoine, C. (2016). *Citizen data science augments data discovery and simplifies data science.* https://www.gartner.com/en/documents/3534848. Accessed 29 July 2023.

TDSP. (n.d.). *What is the team data science process?* https://learn.microsoft.com/en-us/azure/architecture/data-science-process/overview. Accessed 29 July 2023.

Chapter 2
Problem Definition

Before collecting data or thinking about which analytic technique to use, it is critical to understand the business problem. While it may seem evident that the business problem should be clearly stated, it is frequently reported that the biggest reason for failures of analytics projects is a poor understanding of the problem. Moreover, defining a problem that will add value to an organization must be preceded by a good knowledge of the business.[1] This phase of the analytics process includes understanding the business, identifying the stakeholders, recognizing the problem type, and framing the questions. There is a temptation to take a request from a client or manager without careful thought and to go directly to gathering and analyzing data. If this path is taken, the most important questions will frequently be unanswered, wasting time and resources.

The benefits of a careful effort to understand the business and define the problem include:

- Increasing the efficiency of analyst's work because fewer dead ends are encountered. Data collection will be more focused on getting the correct data. Rather than just getting all the data that can be found, a more directed search occurs. Clearly stating which data is needed also provides a more substantial justification for asking the client for additional data in consulting situations.
- Focusing the project on adding value. With the proper problem structure obtaining a helpful solution is more likely. Scattershot model building is expected to lead to poor predictions and little guidance on what is wrong and how to improve the model.
- Avoiding two well-known traps:

  - Providing already known information (e.g., the client says, "Tell me something I don't already know").

---

[1] The word business is used collectively to include non-profit organizations, governments, health-care providers, educational institutions, etc., that can use predictive analytics.

– Providing results that are so unusual as to be unbelievable (e.g., a manager says, "Something is wrong with your analysis; your results strain credibility").

2.1 Expert Views on Problem Definition

Defining the problem is an essential task in any analytics project. These quotes from various experts attest to the importance of this phase of the analytics process.

- "The difference between great and mediocre data science is not about math or engineering: it is about asking the right question(s)… No amount of technical competence or statistical rigor can make up for having solved a useless problem" (Cady, 2017).
- "If I had one hour to save the world, I would spend fifty-five minutes defining the problem and only five minutes finding the solution" (attributed to Einstein (n.d.)).
- "The most serious mistakes are not being made as a result of wrong answers. The truly dangerous thing is asking the wrong questions" (Drucker, 2015).
- "Successful analytic teams spend more time understanding the business problem and less time wading through lakes of data" (Taylor, 2017).
- "What we are finding is that in a lot of companies, there are great data scientists and great businesspeople, but what is missing is businesspeople who know enough data analytics to say, 'Here is the problem I would like you to help me with.' And then, they can take the outcome from the data scientists and see how they can best leverage it." (Ittner, 2019).

2.2 A Telecom Problem Definition Example

One typical application of analytics is predicting and understanding churn (canceling a contract or switching to a different provider) for telecommunications companies. Subscribers to cell phone plans are notorious for canceling contracts and signing up with another carrier. People churn because they are unhappy with the service, want a new phone, or want to get a better price from another provider. Churn can be costly for the provider, and it is usually far cheaper for the provider to take steps to keep a customer than to find a new one. The following case study demonstrates the importance of delving deeper into a situation rather than focusing on the approach asked for by the client.[2]

A large cell phone carrier in the United States was experiencing a high churn and called an analyst to help with this problem. The carrier wanted the business analyst

---

[2]This is based on a workshop presentation by the TMA consulting group. https://the-modeling-agency.com

to find a way to predict which customers were most likely to leave the company and to determine what might be done to prevent those customers from leaving. This seemed like a reasonable request that could be addressed with analytics. The cell phone company had customer data, including a historical database of customers who stayed and left. This type of data can be used to develop a predictive model, but the analyst chose to investigate the problem before building a predictive model.

The analyst found that much of the churn occurred right around the time when a customer's 2-year contract was up. The analyst asked managers in the company for a detailed explanation of how customers were contacted, how they were asked to sign up for a new contract, and what sort of communications were being sent to the customers. It turned out that about two months before a customer's agreement was up, the company sent a reminder letter saying that the contract was coming up for renewal and asking that the customer sign up for another two years.

Imagine you are a consumer with a cell phone company, and you get a reminder letter from the company saying that your contract ends in 60 days. What behavior does this trigger? For many customers, it got them thinking about shopping around. It turned out that the efforts to get customers to sign up for another contract motivated many of them to leave the company. The company's communication triggered the churn, and when these communications about renewing contracts were stopped, the churn was reduced.

If the analyst had just gone directly to analyze the data, the root cause might not have been identified. Management wanted a predictive churn model, but understanding the real problem was the key in this situation.

2.3   Defining the Analytics Problem

Several tasks can be performed to develop a good problem definition:

- Determine the ultimate business objectives. Predictive analytics are used to create value for an organization. In business situations, value-related goals might be to gain new customers, increase current customers' loyalty, increase the efficiency of key processes, or reduce monetary loss due to fraud.
- Translate the business objectives into metrics. Such metrics can be used to assess the performance of developed models that result from the analysis. For example, if customer loyalty is a problem, the objective could be to identify those customers likely to stop using the company's products or services. This objective might be further refined by stating a needed lead time before customer loss so that remedial action could be taken.
- Identify stakeholders. Stakeholders are the people in the organization who care about the problem; these people can be part of the solution, potentially derail the solution, provide the resources needed to arrive at a solution or be able to act on the results. Stakeholders can include IT, executives in various departments (mar-

keting, accounting, finance, etc.), those expected to use the resulting model, and those affected in some way by the project.

- <u>Develop a project plan.</u> Developing a plan involves creating a list of the project stages, required resources, risks, time duration, contingencies, and deployment plans.
- <u>Carefully frame the problem.</u> How a situation is framed can lead to quite different approaches to finding a solution.

Framing refers to describing and interpreting a situation. It focuses attention. The problem frame helps define the importance of a problem and sets the direction for solving it. How a problem or question is framed can lead to quite different answers. For example, these two questions ask about the number 10 in two ways, which will lead to different responses:

What is the sum of 5 plus 5?
What two numbers add up to 10?

Similarly, different frames will lead to different analytics approaches. For example, an insurance company may want to reduce the cost of fraudulent claims. Analytics could address this problem, but framing the situation will lead to different target variables and analytic approaches. Consider the following different frames that could be associated with reducing fraud at an insurance provider:

- Identify customers that had the highest propensity to commit fraud.
- Identify customers most likely to commit fraud in the next 6 months.
- Identify customers who are likely to commit fraud when applying for a policy.
- Identify fraudulent claims where there is the highest likelihood of recovering monies.
- Identify fraudulent claims with the most significant potential dollar amount of fraud.
- Identify fraudulent claims that maximize the hourly return of investigators assigned to deal with potential fraud.

Sometimes, the first attempt at a problem statement focuses on symptoms, which may not address the root problem. For example, a sign might be: "We are losing market share rapidly." The problem definition might be: "How do we regain market share?" (Hauch, 2018). But this may not direct the analyst team to the real issue.

One suggestion to get at the root of a problem through framing is to use "The 40-20-10-5 rule." You can apply this rule through four basic steps for complex problems. State your problem in 40 words. Cut it down to 20, then to 10, and end up with a five-word problem statement. If you cannot keep the problem definition simple, you have not reached its roots yet.[3]

Another data scientist summarized key considerations as a guide to problem framing when proposing a predictive analytics model from attending a Google training session (Arnuld, 2020).

---

[3] https://probsolvinglburkholder.weebly.com/creativity.html

- See if you can solve the problem without using predictive analytics. Sometimes a simple heuristic is good enough.
- Be clear about what you want and state it without using predictive analytics.
- Write your desired outcome in simple English.
- What do you expect the model to output?
- What predictive analytics problem are you solving, e.g., if supervised, then what kind: classification or regression? If classification, then what kind? Binary or multiclass?
- What is a successful model in this case?
- What is a failed model in this case?
- Quantify your success. How will you measure it in your model?
- Try to keep the model simple. Complex models are slower to train and difficult to understand.
- Build a simple model and deploy it. The results from deploying the simple model can justify a more complex model.

An example of solving a more straightforward problem was described by Flinchbaugh (2009). Engineers can sometimes make things more complicated than necessary. Consider the example of the NASA space pen. During the 1960s, NASA launched a program to develop a pen that would write in zero gravity. The NASA program was costly. The Soviets used the much cheaper already-invented pencil. Apparently, the American engineers worked on the problem, "How can we make a pen that will work in zero gravity?" while the Soviets asked, "What will write in zero gravity?"

2.4   Structured Versus Unstructured Problems

The types of problems that could be addressed with analytics are wide-ranging:

- How can the Red Cross increase blood donations?
- Should a consumer products company increase ad spending for a specific brand or lower the price to stimulate sales?
- How do you increase the number of people who agree to be organ donors?
- How can a university improve its student retention rate?
- How much of a discount should a resort offer for booking tickets online 6 months early?
- Should a hotel offer a "surprise discount" on rooms at check-in, or should they offer it as people leave?
- How can a casino increase the lifetime value of a customer?
- Should a particular person be approved for a loan?
- Is this credit card transaction fraudulent?

While these may seem like precise questions, attempting to structure the question to guide an analytics project would likely reveal that the questions are poorly structured.

Many challenging analytics problems are unstructured and complex. Professionals are rewarded for their ability to solve difficult problems, not simply following rote procedures or relaying memorized information. When faced with an unstructured problem, it is even more critical to define it carefully. Based on Jonassen (1997) the following lists some characteristics that distinguish well-structured versus unstructured problems.

Characteristics of Well-Structured Problems
- All elements are clearly presented.
- A possible solution is available.
- There exist a limited number of rules and principles that can be applied.
- The concepts and rules are in a definable knowledge domain.
- There are well-defined constraints.
- There exist correct, convergent answers.
- There is a preferred, prescribed solution process.

Characteristics of Unstructured Problems
- Goals are vaguely defined or unclear.
- There are multiple criteria for evaluating solutions.
- There are no general rules or principles to guide a solution.
- Relationships among concepts, rules, and principles are inconsistent.
- Constraints are unstated or unclear.
- There are multiple (or no) solution paths.
- There are no clear means for determining appropriate actions.

2.5   Getting Started with Defining the Problem

Because of the critical importance of problem definition, many tools and strategies have been developed to get started. A few of these are listed here.

Right to Left Thinking
One approach that can be very helpful is to turn the business objective into a decision to be made. For example, if the problem is that a company needs more customers, this could lead to a goal of getting new customers or working on keeping current customers. These goals will lead to quite different data needs and analyses. If the goal is keeping existing customers, then a model that predicts which specific customers are likely to leave can lead to actions to reduce the probability of those customers leaving. If, on the other hand, the goal is to get new customers, then the model can be used to indicate which individuals are most likely to respond favorably to an offer. So, using right-to-left thinking, the process would begin at the "right side" with an operational decision and work backward to specify the data need, the risks, the buy-in required, and so on.

Reversing the Problem
Identify the problem or challenge and write it down. Brainstorm the reverse problem to generate reverse solution ideas. Allow the brainstorming to flow freely. Do not reject anything at this stage.

Instead of asking, "How do I solve or prevent this problem?" ask, "How could I cause the problem?" Instead of asking, "How do I achieve these results?" ask, "How could I achieve the opposite effect?"

Once you have brainstormed all the ideas to solve the reverse problem, reverse these into solutions for the original problem or challenge. Evaluate these ideas to determine if a potential solution is suggested or at least the attributes of a solution (Puccio, 2016).

Opening Up by Asking "Whys"
Consider the following example of opening up the problem by asking a series of "whys":

If I asked you to build a bridge for me, you could go off and build a bridge. Or you could come back to me with another question: "Why do you need a bridge?" I might need a bridge to get to the other side of a river. Aha! This response opens up the frame of possible solutions. There are many ways to get across a river besides using a bridge. You could dig a tunnel, take a ferry, paddle a canoe, use a zip line, or fly a hot-air balloon, to name a few.

You can open the frame even further by asking why I want to get to the other side of the river. Imagine that I told you that I work on the other side. This information broadens the range of possible solutions even more. There are probably viable ways to earn a living without ever going across the river (Seelig, 2013).

Challenging Assumptions
No matter how simple it may be, every problem is attached to a lengthy list of assumptions. Many of these assumptions may need to be revised and could make your problem statement inadequate or even misguided.

The first step is making assumptions explicit. Write a list of as many assumptions as possible – especially those that may seem the most obvious and 'untouchable.' That brings more clarity to the problem at hand. It would help if you learned to think like a philosopher.

But go further and test each assumption for validity: think in ways that might not be valid and their consequences. What you will find may surprise you: that many of those flawed assumptions are self-imposed – with just a bit of scrutiny, you can safely drop them. Be a skeptic (Clissold, 2021).

Consider the example of the mathematician Abraham Wald and the British Air Ministry.[4] During World War II, the British Air Ministry engaged Mr. Wald to study the damage suffered by airplanes involved in combat missions and use those results to recommend appropriate places for armor reinforcement. All returning planes were assessed for damage, and the data was collected and analyzed. Patterns of

---

[4] https://www.outlookindia.com/website/story/a-world-war-ii-puzzle/294404

damage soon became apparent, and the officers of the Royal Air Force concluded that the planes should be reinforced based on those patterns.

Mr. Wald took a different approach. He reasoned that the damaged planes which made it back were not the real problem. Additional armor was needed most on planes that did not make it back. While Wald didn't know where the damage was on the lost aircraft, he could reason that it differed from the damage to planes that safely returned. His task was to see the problem, which was not easily observed (Ellenberg, 2016).

Chunking

Chunking[5] up is about taking a broader view. Helicopter up to 30,000 feet. Survey the landscape to see the entire system. Ask 'Why' things happen to find a higher-level purpose. Ask 'What is this an instance of' to find a more general classification. Use inductive reasoning to go from specific details to general theories and explanations.

Chunking down is about going into detail to find minor, more specific system elements. Ask 'How' things happen to find lower-level detail. Ask 'What, specifically' to probe for more information. Ask "Give me an example" to get specific instances of a class. Use deductive reasoning to go from general theories and ideas to cases.

Chunking up and down go well together to look differently at the same situation. Chunk up from the existing situation to find a general or broader view. Then chunk down somewhere else.

2.6 Summary

Preparing a practical, actionable definition of the problem before developing a machine learning application is universally recognized as necessary. However, specifying exactly how to create a problem definition can be complicated in a complex process. As with many complex concepts, it is likely that "you will know it when you see it." But getting to that point requires creativity and hard work.

The effort expended in problem definition will improve the chances that the results of an analytics project are implemented and achieve the desired results. Several methods of forcing the analysts or an analytics team to dig deeper into problems were discussed. Using one or more of these tools (or other similar tools) will help prevent the tendency to obtain data and run the computer immediately. The errors that can result from superficial efforts to define the problem can make using analytics a costly and frustrating experience.

---

[5] http://changingminds.org/:   http://changingminds.org/techniques/questioning/chunking_questions.htm

References

Arnuld. (2020). *Some key things I learned from Google's introduction to machine learning problem framing MOOC.* https://www.linkedin.com/pulse/some-key-things-i-learned-from-googles-introduction-machine-on-data. Accessed 29 July 2023.

Cady, F. (2017). *The data science handbook.* Wiley.

Clissold, R. (2021). *How to solve a problem.* https://www.wikihow.com/Solve-a-Problem. Accessed 29 July 2023.

Drucker, P. (2015). *Peter Drucker on asking the wrong questions.* https://hac.bard.edu/amor-mundi/peter-drucker-on-asking-the-wrong-questions-2015-09-29. Accessed 29 July 2023.

Einstein, A. (n.d.). https://www.azquotes.com/author/4399-Albert_Einstein/tag/problem-solving. Accessed 29 July 2023.

Ellenberg, J. (2016). *Abraham Wald and the missing bullet holes.* https://medium.com/@penguin-press/an-excerpt-from-how-not-to-be-wrong-by-jordan-ellenberg-664e708cfc3d. Accessed 29 July 2023.

Flinchbaugh, J. (2009). *Leading lean: Solve the right problem.* https://www.assemblymag.com/articles/86658-leading-lean-solve-the-right-problem. Accessed 29 July 2023.

Hauch, B. (2018). *Finding the true north of a problem: Problem-framing principles for design-led innovation.* https://medium.com/nyc-design/finding-the-true-north-of-a-problem-problem-framing-principles-for-design-led-innovation-b0c7620317bf. Accessed 29 July 2023.

Ittner, C. (2019). *How digital technology is transforming cost management.* https://knowledge.wharton.upenn.edu, https://knowledge.wharton.upenn.edu/article/cost-management-in-the-digital-age/. Accessed 29 July 2023.

Jonassen, D. (1997). Instructional design models for well-structured and ill-structured problem-solving learning outcomes. *Educational Technology Research and Development., 45*, 65–94.

Puccio, G. (2016). *How to use reverse brainstorming to generate fresh ideas.* https://www.wondri-umdaily.com/reverse-brainstorming/ Accessed 29 July 2023.

Seelig, T. (2013). *Shift your lens: The power of re-framing problems.* http://stvp.stanford.edu/blog/?p=6435. Accessed 29 July, 2023.

Taylor, J. (2017). *Bringing business clarity to CRISP-DM.* https://www.kdnuggets.com/2017/01/business-clarity-crisp-dm.html. Accessed 29 July 2023.

Chapter 3
Introduction to KNIME

KNIME is a comprehensive analytics and data mining tool that uses an intuitive drag-and-drop workflow canvas. While KNIME can be (and is) used by professional data analysts, it is an excellent "low or no code" platform for predictive analytics and data mining. KNIME workflows provide a graphic representation of the steps taken in analysis, making the analysis self-documenting. This documentation makes it easy to reproduce analyses and communicate methods and results to others.[1]

This chapter introduces KNIME with demonstrations of some of its features and includes references to various free resources, including tutorials and documentation. A step-by-step example of using KNIME demonstrates how KNIME can be used to analyze a data set.

KNIME was developed at the University of Konstanz in Germany beginning in 2004, with the first release in 2006. As noted on its website, KNIME is committed to open source, and the KNIME Analytics Platform is not a cut-down version.

3.1 KNIME Features

There are several features that make KNIME stand out from its competitors: KNIME:

- Is free to use on your machine.
- Runs on Windows, Mac, and Linux machines.
- Has over 4000 nodes for data source connections, transformations, machine learning, and visualization.

---

[1] If an error is encountered when running KNIME workflows supplied by others, it may be necessary to load nodes not included in the default installation. You will be prompted to load the nodes if the error occurs.

- Is fully extensible and can run R or Python to extend its capabilities. In addition, many of the capabilities of two other analytic platforms, H2O, and WEKA, are also integrated into KNIME and work seamlessly in the drag-and-drop workflow.
- Can read various data file types, including CSV, Excel, JSON, XML unstructured documents, images, or audio files.
- Can export results to Excel, Tableau, Spotfire, Power BI, and other reporting platforms.
- Has a large active community of users who can answer questions and help.
- Provides many ready-to-use workflows that can be easily installed in your work environment by dragging and dropping from the KNIME site.
- Offers extensive documentation, learning modules, videos, training, and user events.

A paid, commercial server version of KNIME is needed for deploying KNIME models on the web, but this is not required for desktop applications. Everything else is available with the open-source version.

3.2   The KNIME Workbench

The Workbench appears after launching KNIME. The main sections of the KNIME Workbench are shown in Fig. 3.1; a brief description of each is provided below. For more detail, see the KNIME Workbench Guide.[2]

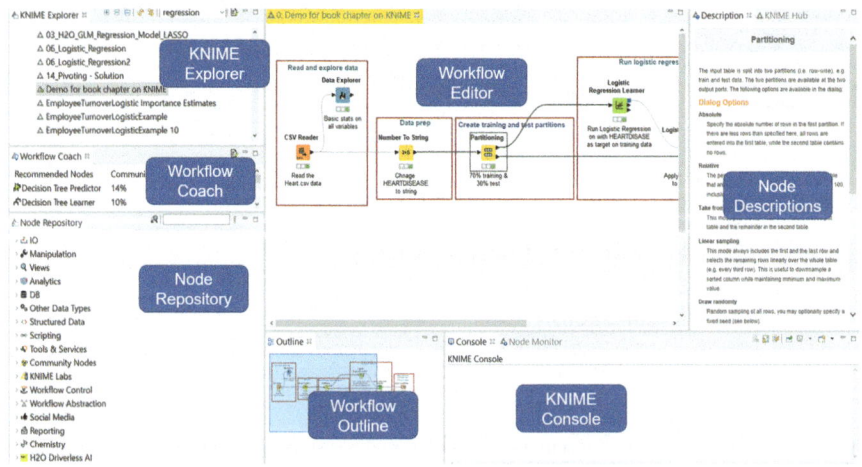

Fig. 3.1 The KNIME Workbench

---

[2] https://docs.knime.com/2018-12/analytics_platform_workbench_guide/index.html#workspaces

Elements of the KNIME Workbench

KNIME Explorer. This links to the available workflows on your machine and those available from KNIME servers, including workflow examples from KNIME and the KNIME community.

Workflow Coach. This is a handy tool to help you build an analytic workflow. The Coach suggests connecting nodes and processes to any node selected in your workflow. The suggestions by the Coach are based on actual workflows constructed by users in the KNIME Community.

Node Repository. Listed in this area are the analysis nodes installed on your machine. Nodes are available to read and write files, explore and transform data, run basic and advanced analytics, and create visualizations. A core set of nodes is included when you install KNIME, but thousands of additional nodes are available and easily installed. The nodes are organized by categories, but you can also use the search box on the top of the node repository to find nodes.

Workflow Editor. This is the canvas for editing the currently active workflow. The workflow is created by dragging nodes from the Node Repository and linking them interactively.

Outline. This area shows a small overview of the current workflow.

Console. This shows the processing taking place when executing a workflow. It also provides warnings and error messages.

Node Descriptions. For each node selected in a workflow, a detailed description is provided, including the general function of the node, available settings, and input and output ports.

3.3   Learning to Use KNIME

Learning to use KNIME takes some time, but extensive, free written and video resources are available. A "Getting Started Guide" is available from KNIME.[3]

Four levels of self-paced courses with exercises are online and free.[4]

Level 1 courses

- KNIME Analytics Platform for Data Scientists: Basics
- KNIME Analytics Platform for Data Wranglers: Basics

Level 2 courses

- KNIME Analytics Platform for Data Scientists: Advanced
- KNIME Analytics Platform for Data Wranglers: Advanced

Level 3 course

- KNIME Server Course: Productionizing and Collaboration

---

[3] https://www.knime.com/getting-started-guide

[4] https://www.knime.com/knime-courses

Level 4 courses

- Introduction to Big Data with KNIME Analytics Platform
- Introduction to Text Processing

KNIME also provides documentation,[5] which can be read online or downloaded. Individual documents for specific topics are available as follows:

- KNIME QuickStart Guide
- KNIME Analytics Platform Installation Guide
- KNIME Workbench Guide
- KNIME Extensions and Integrations Guide
- KNIME Flow Control Guide
- KNIME Components Guide
- KNIME Integrated Deployment Guide
- KNIME File Handling Guide

A free "boot camp" for KNIME is available from UDEMY[6] with 50 instructional videos running over four hours. The course starts with installation and setup and proceeds to demonstrate practical applications in machine learning.

3.4   KNIME Extensions and Integrations

In addition to the nodes that are developed and maintained by KNIME, there are many nodes that community developers and KNIME Partners have developed. The KNIME Community Extensions offer nodes for image processing, information retrieval, and many others that are created and maintained by community developers. Trusted Extensions provide the following guarantees to users:[7]

- Backward compatibility, i.e., existing workflows will continue working with newer KNIME versions.
- Compliance with the KNIME usage model and quality standards.
- Support (via forum) in case of problems.
- Maintenance for the previous two KNIME versions.

As mentioned earlier, one powerful feature of KNIME is its integration with open-source tools. Examples are listed below.

- R and Python scripts can be run seamlessly in a KNIME workflow. Code can also be run from Jupyter Notebooks.
- H20 provides high-performance algorithms such as Random Forests, Gradient Boosted Trees, and many others. The H20 open-source tools can be integrated

---

[5] https://docs.knime.com/

[6] https://www.udemy.com/course/knime-for-beginners-a-practical-insight/

[7] https://www.knime.com/trusted-community-contributions

into a KNIME workflow. One feature of H20 is an available "AutoML Learner," which runs and compares several predictor models' performance using optimum parameter settings.

- KNIME provides integration with the Python Keras Deep Learning framework. Keras (which makes TensorFlow more approachable) contains several implementations of neural networks, including convolutional and recurrent nets, which can be used for applications such as image recognition.

3.5   Data Types in KNIME

Data tables in KNIME support several different data types, including Double (D), Integer (I), String (S), Date or Time, and Boolean (B). The data type is indicated in the header row of each variable. Double values are floating point numerical values with fractional parts. Double floating-point variables allow greater precision by using 64 bits. Integer variables store signed whole numbers. String variables (alphanumeric or character variables) may include letters, numbers, or symbols. Date variables are strings formatted as yyyy-MM-dd and Time variables formatted as HH:mm:ss. Boolean variables represent True/False or 0/1.

3.6   Example: Predicting Heart Disease with KNIME

The following example illustrates how KNIME can be used to build a model to predict the presence of heart disease. Do not be concerned if some of the details are unfamiliar at this point; these details will be covered in later chapters of the book when this example is revisited.

Mortality rates for cardiovascular diseases are high and increasing in many regions of the world. In the United States, for example, one in four deaths annually is from heart disease. If the presence of heart disease can be predicted using a physician's examination and laboratory tests, this would be a valuable diagnostic tool. Such a prediction is difficult because there are several contributory variables, including diabetes, high blood pressure, high cholesterol, abnormal pulse rate, and other factors.

A data set from the Cleveland Clinic, a leading cardiovascular research hospital, is available from the UCI Machine Learning Repository.[8] The data set has 297[9] rows, 13 predictor values, and one target (heart disease versus no heart disease). Since the target variable has just two levels, logistic regression was run in

---

[8] https://archive.ics.uci.edu/dataset/45/heart+disease
[9] Six rows were dropped due to missing values.

KNIME. (Details on logistic regression and other models are provided in later chapters).

The workflow for this example is shown in Fig. 3.2.

The first step in the Fig. 3.2 workflow reads the heart disease data with the CSV Reader node. The node was configured as shown in Fig. 3.3, and a preview of the data is shown in Fig. 3.4.

Note that the data had been saved earlier so the file was saved "Relative to" the "Current workflow data area." This conveniently bundles the data with the workflow to be stored, shared with others, and run if the users have KNIME installed on their machines.

Good practice in building machine learning models is to use part of the data for training and the rest for testing the model. Accordingly, the KNIME node "Partitioning" was used to create training and test data subsets. Eighty percent of the sample was used for training and 20% for testing (Fig. 3.5). The 80/20 ratio is typical but not the only split that can be used.

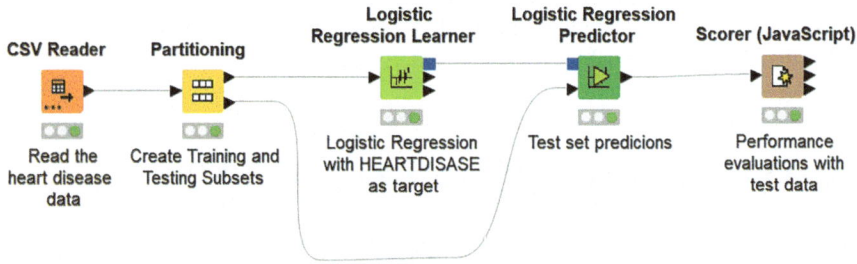

Fig. 3.2 Example of a KNIME workflow for predicting heart disease

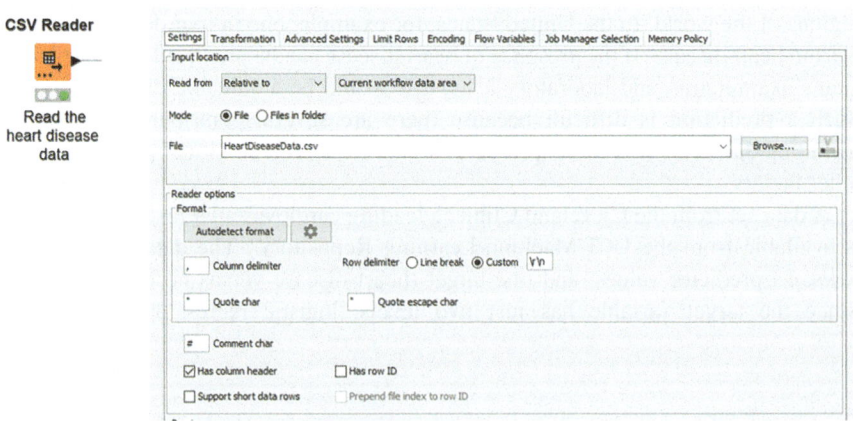

Fig. 3.3 Configuring the CSV Reader node

File Edit Hilite Navigation View

Table "default" - Rows: 297 Spec - Columns: 14 Properties Flow Variables

Row ID	Age	Sex	Chestpain	Resting...	Cholest...	Fasting...	Resting_ECG	Max_h...	Exercis...	ST_dep...	Peak_s...	Num_,...	Thalass...	Heart_...
Row0	63	male	typical_angina	145	233	TRUE	left_ventricular_hyperthophy	150	no	2.3	downsloping	None	fixed_defect	not_present
Row1	37	male	non_angina...	130	250	FALSE	normal	187	no	3.5	downsloping	None	normal	not_present
Row2	41	female	atypical_ang...	130	204	FALSE	left_ventricular_hyperthophy	172	no	1.4	upsloping	None	normal	not_present
Row3	56	male	atypical_ang...	120	236	FALSE	normal	178	no	0.8	upsloping	None	normal	not_present
Row4	57	female	asymptomatic	120	354	FALSE	normal	163	yes	0.6	upsloping	None	normal	not_present
Row5	53	male	asymptomatic	140	203	TRUE	left_ventricular_hyperthophy	155	yes	3.1	downsloping	None	reversible_d...	present
Row6	57	male	asymptomatic	140	192	FALSE	normal	148	no	0.4	flat	None	fixed_defect	not_present
Row7	56	female	atypical_ang...	140	294	FALSE	left_ventricular_hyperthophy	153	no	1.3	flat	None	normal	not_present
Row8	44	male	atypical_ang...	120	263	FALSE	normal	173	no	0	upsloping	None	reversible_d...	not_present
Row9	52	male	non_angina...	172	199	TRUE	normal	162	no	0.5	upsloping	None	reversible_d...	not_present

Fig. 3.4 The first ten rows of the Heart Disease data

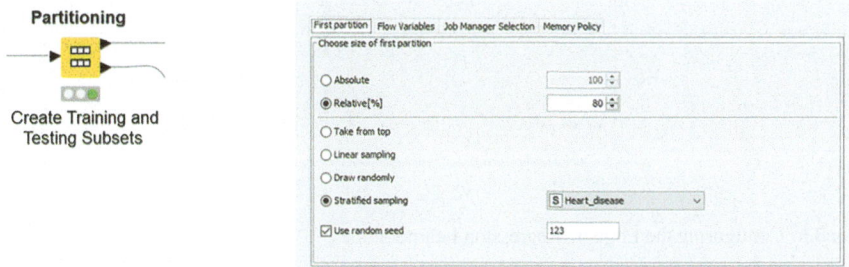

Fig. 3.5 Create training and test data sets

The data was split randomly; a seed was set to 123 to make the randomization repeatable. Also, the random selection was conducted with stratified sampling so that the ratio of "heart disease" versus "no heart disease" would be equal (or nearly equal) in both the training and test subsets. The Partitioning node has two outputs: the upper output is the training data, and the lower output is the test data.

The training portion of the data from the upper output port of the Partitioning node was linked to the input of the Logistic Regression Learner node (Fig. 3.6). The target variable was set to Heart_disease, and the remaining variables were included as predictors. For each categorical variable, the Learner node appends columns equal to the number of levels in the variable. These indicators or "dummy" variables are used as predictors instead of the original categorical variables. (Additional ways to deal with categorical predictors will be discussed in the ordinary least squares regression chapter).

The Logistic Regression Predictor node (Fig. 3.7) has two inputs: the square input port takes the model built in the Learner node, and the lower arrow input port takes the test data from the Partitioning node. The Logistic Regression Predictor uses these inputs to create predictions for the test data, which is output in the single arrow on the node's right.

There are three outputs from the Logistic Regression Learning: the upper output (shown as a dark square) contains the logistic model, and the center output contains the coefficients and statistical tests. The lower output arrow has information about the modeling process, including the number of iterations required and the log-likelihood.

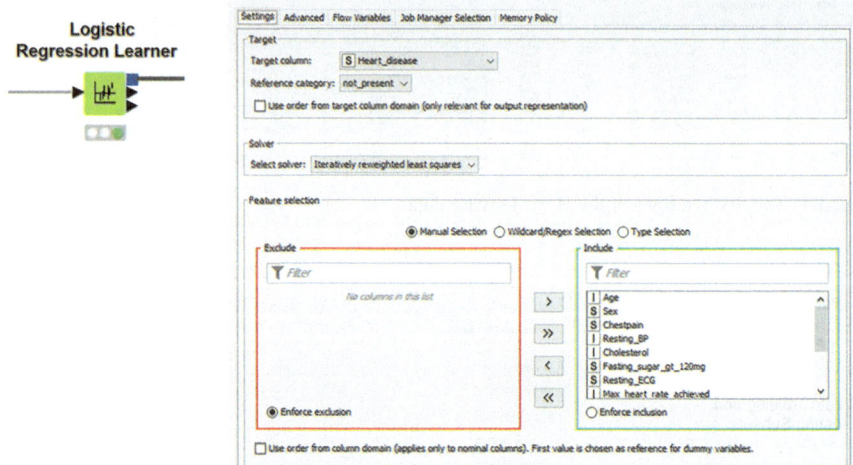

Fig. 3.6 Configuring the Logistic Regression Learner node

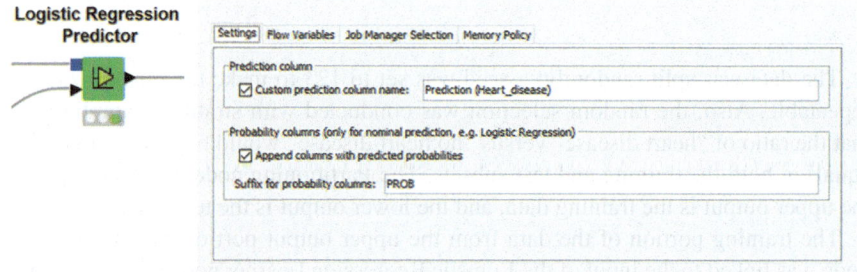

Fig. 3.7 Configuring the Logistic Regression Predictor node

The upper output from the Learner node becomes the upper input to the Logistic Regression Predictor. The lower input is the test data. Note that the model was created using the training data, but the predictions are made with the test data. This provides an independent test of the model's performance on unseen data. Three columns are appended to the test data: the predicted probability that heart disease is "present", the predicted probability that heart disease is "not present", and the binary prediction of heart disease "present" or "not present" using a threshold of 0.5. The first 20 rows and the right-most four columns of the output data table from the Logistic Regression Predictor node are shown in Fig. 3.8. The left column contains the actual condition of the patient, and the right column contains the predicted condition. The two central columns contain the estimated probabilities. If the probability of disease was greater than 0.50, then the prediction was "present" otherwise

S Heart_...	D P (Hear...	D P (Hear...	S Predicti...
not_present	0.958	0.042	not_present
not_present	0.962	0.038	not_present
not_present	0.963	0.037	not_present
not_present	0.798	0.202	not_present
present	0.879	0.121	not_present
not_present	0.994	0.006	not_present
not_present	0.931	0.069	not_present
not_present	0.984	0.016	not_present
present	0.969	0.031	not_present
present	0.523	0.477	not_present
not_present	0.948	0.052	not_present
not_present	0.991	0.009	not_present
not_present	0.818	0.182	not_present
present	0.199	0.801	present
present	0.129	0.871	present
not_present	0.417	0.583	present
not_present	0.763	0.237	not_present
not_present	0.874	0.126	not_present
present	0.344	0.656	present
not_present	0.991	0.009	not_present

Fig. 3.8 The first 20 rows of the Logistic Regression Predictor output

the prediction is "not present." The results shown in Fig. 3.8 indicate some correct and some incorrect predictions. The actual condition is "not present" in row five, yet the model predicted "not present."

The Scorer node in the workflow shown in Fig. 3.9 is used to evaluate the performance of models where the target variable is binary, as is the case in this example. The input to this node is a data table with the actual and predicted values for the test data set. A cross-tabulation called a "confusion matrix" is created using the actual and predicted values for the presence of heart disease. The upper output port of the Scorer contains the confusion matrix, and the lower two ports contain performance statistics related to the model predictions. These will be discussed in the chapter on logistic regression.

The confusion matrix from the Scorer node indicated that the logistic regression model correctly predicted 80% of the cases, which is most likely not sufficiently accurate for clinical diagnoses. A different model with more variables might be developed to improve accuracy. Especially of concern in the results in the confusion matrix were the 11 cases where heart disease was "present," but the model predicted "not present."

Scorer (JavaScript)

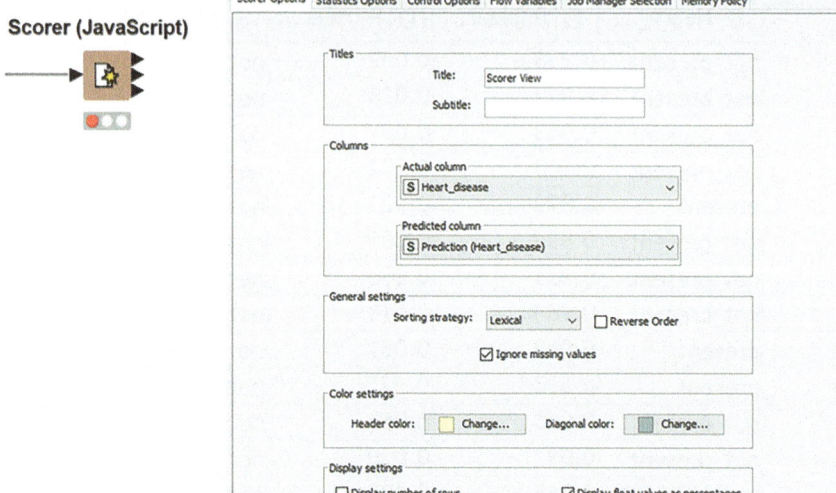

Fig. 3.9 Configuring the Scorer node

3.7   Example: Preparation of Hospital Data Using KNIME

The length of stay for hospital patients is an important indicator of the efficiency of care. Extended stays can increase the likelihood of infection and complications, and extended stays in the hospital negatively impact a patient's experience and result in higher healthcare costs.

This example uses a 2015 data set with over 2.3 million patient hospitalizations in New York State.[10] Thirty-four variables are included in the data set, but this analysis will focus on the variables in Table 3.1.

The workflow for the analysis is shown in Fig. 3.10. Node descriptions are shown in Table 3.2.

The following steps would be used to run the workflow:[11]

1. Make sure the workflow is in a hard-drive folder.
2. Launch KNIME.

---

[10] https://health.data.ny.gov/Health/Hospital-Inpatient-Discharges-SPARCS-De-Identified/82xm-y6g8. Accessed 28 July 2023.

[11] Many nodes in KNIME use "regex" for searching and transforming variables. Regex (which stands for "regular expressions") uses a sequence of special characters to select variables and perform operations. Examples of regex expressions are provided in Appendix 2 of this chapter.

3. Click on File> Import KNIME Workflow.[12]
4. In the Select file area: Browse to the file WorkflowForFigure3.10.knwf file and
   select it.
5. Click Open, then Finish.
6. Locate the selected workflow in the KNIME Explorer and double-click on the
   workflow file.
7. Execute the nodes in order from left to right.

Table 3.1 Variables in the hospital discharge data set

Variable	Description
Age Group	0–17, 18–29, 30–49, 50–69, 70 or Older
Gender	M, F, U (male, female, unknown)
Length of Stay	In days
Type of admission	Elective, Urgent
Severity of the illness description	Minor, Moderate, Major, Extreme
Total Charges	The amount billed by the hospital
Total Cost	The expense incurred by the hospital

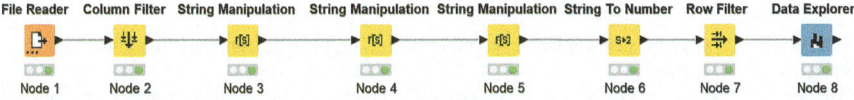

Fig. 3.10 Workflow for analysis of hospital data subset

Table 3.2 Node descriptions for the Fig. 3.10 workflow

Nodes	Descriptions	Input and output ports
Node 1 File Reader	Read file: Hospital_Inpatient_Discharges	Output port: Data table with Hospital_Inpatient_Discharges with 2,346,931 rows and 34 columns

Node 2
Column Filter
	Place the following variables in the Include area:
	    Age Group
	    Gender
	    Length of Stay
	    Type of Admission
	    APR Severity of Illness Description
	    Total Charges
	    Total Costs
	Put other variables in the Exclude area
		Input port: Data table from the Node 1 Output port
		Output port: Selected columns of the data table with Hospital Inpatient Discharges with 2,346,931 rows and 7 columns

(continued)

[12] The workflow files are available online at https://tinyurl.com/KNIMEWorkflows

Table 3.2 (continued)

Nodes	Descriptions	Input and output ports
Node 3 String Manipulation	String manipulation to remove the comma from the variable Total Charges In the Expression area: removeChars($Total Charges$,",") Select: Replace Column Enter name: Total Charges	Input port: Data table from the Node 2 Output port Output port: Modified data table: Removed "," from Total Charges
Node 4 String Manipulation	String manipulation to remove the comma from the variable Total Costs In the Expression area: removeChars($Total Costs$,",") Select: Replace Column Enter name: Total Costs	Input port: Data table from the Node 3 Output port Output port: Modified data table: Removed "," from Total Costs
Node 5 String Manipulation	String manipulation to remove the plus sign from the variable Length of Stay In the Expression area: removeChars($Length of Stay$,"+") Select: Replace Column Enter name: Length of Stay	Input port: Data table from the Node 4 Output port Output port: Modified data table: Removed "+" from Length of Stay
Node 6 String to Number	Change string variables to numeric Settings: Place the following variables in the Include area: Length of Stay Total Charges Total Costs	Input port: Data table from the Node 5 Output port Output port: Modified data table with Length of Stay, Total Charges, and Total Costs changed from string to numeric
Node 7 Row Filter	This node removes the cases where Gender is U (unknown) Filter Criteria: In the left box, select Exclude rows by attribute value Column to test: Gender Matching criteria: Check to use pattern matching; enter the letter U	Input port: Data table from Node 6 Output port Output port: Modified data cases with Gender = "U" removed; 2,346,883 rows by 7 columns
Node 8 Data Explorer	Option Columns Check Show most frequent/ infrequent nominal values Check Display Row Id in the Data Preview	Input port: Data table from Node 7 Output port Output port: Same as the input port

Descriptive Statistics from the Data Explorer

The output from the Data Explorer node for the continuous variables is shown in Fig. 3.11, and the output for nominal variables is shown in Fig. 3.12.

Figure 3.11 shows that Total Charges, Total Costs, and Length of Stay are positively skewed. No missing values are observed for any of the continuous variables.

Column	Exclude Column	Minimum	Maximum	Mean	Standard Deviation	Variance	Skewness	Kurtosis	Overall Sum	No. zeros
Length of Stay	☐	1	120	5.476	8.034	64.544	6.469	64.441	12852621	0
Total Charges	☐	0.010	7248390.820	43215.410	80460.751	6473932469.688	14.041	445.589	101421510512.161	0
Total Costs	☐	0	3007712.730	14732.745	28199.134	795191171.527	17.868	767.328	34576029543.785	3

Fig. 3.11 Summary of continuous variables in the hospital data subset

Column	Exclude Column	No. missings	Unique values	Frequency Bar Chart
Age Group	☐	0	5	
Gender	☐	0	2	
Type of Admission	☐	0	6	
APR Severity of Illness Code	☐	0	5	

Fig. 3.12 Summary of categorical variables in the hospital data subset

Figure 3.12 shows 103 missing cases for APR Severity of Illness Description in the data set. The small frequency charts represent the number of cases by value, sorted from the highest number of cases to the smallest number.

3.8   Flow Variables

Flow variables are parameters in KNIME used to pass information and overwrite configuration settings. The input and output ports of nodes are used to transfer values in KNIME data tables. But there can be other variables of interest contained in nodes. These variables are the parameters that control settings for nodes and can be double, integer, or string variables. Flow variables can be used to create loops to iterate over values, automate inputs, make workflows interactive, and perform other tasks.

Flow variable ports are shown as red circles above a node. The flow variable ports are not visible when placed on the workflow canvas. To unhide the flow variable ports, right-click on a node and select "Show Flow Variable Ports."

There are many ways to create and use flow variables. Several basic examples will be shown here, and others will be shown in later chapters in specific applications.

Convert a Table Row into a Flow Variable

The example uses an auto insurance fraud data set available from Kaggle.[13] The data set contains 1000 observations with 38 columns with attributes such as the age of the driver, the state where the auto insurance policy was written, the make, model, and year of the policyholder's automobile, the dollar value of the insurance claim, whether the claim was fraudulent. The Table Row to Variable node converts the first row of each column in a data table into a flow variable. The workflow shown in Fig. 3.13 illustrates an application that selects the state (IL, IN, or OH) with the largest total value of claims and reports the state name and the total value. Node descriptions are shown in Table 3.3.

Export a Configuration from a Node as a Flow Variable

The workflow in Fig. 3.14 demonstrates how to export a configuration from one node to another. The example uses the auto insurance fraud data from the previous example and performs two-sample t-tests using the grouping of fraud = yes vs. fraud = no.

The following null hypotheses are tested:

- There is no difference in the mean value of property claims for observations with fraud = Y versus fraud = N.
- There is no difference in the mean value of injury claims for observations with fraud = Y versus fraud = N.

The group variate (fraud = Y or N) is set for the t-test in Node 2 and then exported as a flow variable to the t-test in Node 3. Descriptions of the nodes are in Table 3.4. Results of the tests:

- The null hypothesis of no difference in property claims is rejected with fraud = Y (mean claim amount = \$8560) versus fraud = N (mean claim amount = \$7019).
- The null hypothesis of no difference in personal injury claims is rejected with fraud = Y (mean claim amount = \$8208) versus fraud = N (mean = claim amount \$7179).

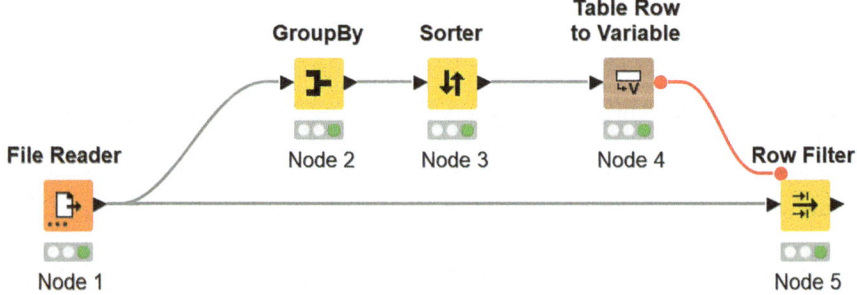

Fig. 3.13 Workflow to demonstrate converting a table row to a variable

---

[13] https://www.kaggle.com/datasets/buntyshah/auto-insurance-claims-data

Table 3.3 Node descriptions for the Fig. 3.13 workflow

Nodes	Descriptions	Input and output ports
Node 1 File Reader	Read file: FraudData.csv	<u>Output port:</u> Data table FraudData with 1000 rows and 38 variables
Node 2 GroupBY	Form groups by the state where the policy was written and obtain the total claim amount for each state Settings: Groups Group column: policy_state Manual Aggregation Column: total_claim_amount Aggregation: Sum	<u>Input port:</u> Data table from the Node 1 Output port <u>Output port:</u> Grouped table with total claim amounts by state with 3 rows and 2 columns with total claim amount for each of three states
Node 3 Sorter	Sorts the rows by the Sum (total_claim_amount) from highest to lowest Sorting Filter: Sum(total_claim_amount) Select: Descending	<u>Input port:</u> Data table from the Node 2 Output port <u>Output port:</u> Sorted table with 3 rows and 2 columns. State with highest total claim amount is placed in the first row
Node 4 Table Row to Variable	Converts the first row of the sorted table (with the highest total claim amount) to a flow variable Settings: Include: policy_state Sum(total_claim_amount)	<u>Input port:</u> Data table from the Node 3 Output port <u>Variable output (red circle):</u> A flow variable with the first row of the sorted table containing the state name and the total value of claims for the state
Node 5 Row Filter	Filter the Fraud data set by the flow variable Filter Criteria: Include rows by attribute value Column to test: policy_state Select: Use pattern matching; click on the V icon and use variable policy_state	<u>Input port:</u> Data table from the Node 1 Output port <u>Flow variable input:</u> A flow variable with the first row of the sorted table (red circle) from Node 4 <u>Output port:</u> Total dollar value claims data for the state with the highest total

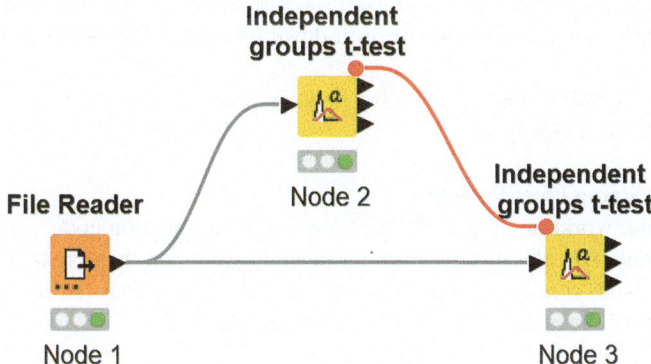

Fig. 3.14 Exporting a configuration using a flow variable

Table 3.4 Node descriptions for the Fig. 3.14 workflow

Nodes	Description	Input and output ports
Node 1 File Reader	Read file: FraudData.csv	Output port: FraudData data table with 1000 rows and 38 variables

Node 2
Independent groups t-test
	Independent two-group t-test of difference mean amounts of property claims
	Settings:
	  Grouping column: fraud-reported
	  Include: property_claim
	  Exclude all other variables
	Flow Variables:
	  In the white space to the right of grouping Column enter GroupCol
		Input port: Data table from the Node 1 Output port
		Flow variable output: GroupCol
		Upper output port: t-test
		Middle Output port: Levene test.[a]
		Lower output port: Descriptive statistics

Node 3
Independent groups t-test
	Independent two-group t-test of difference mean amounts of injury claims.Settings:
	  Grouping column: fraud-reported flow variable from Node 2
	  Include: injury_claim
	  Exclude all other variables
	Flow Variables:
	  In the white space to the right of groupingColumn enter GroupCol
		Input port: Data table from the Node 1 Output port
		Flow variable input: Flow variable GroupCol
		Upper output port: t-test
		Middle Output port: Levene test
		Lower output port: Descriptive statistics

[a]The Levene test assesses whether the variance in the two groups for a t-test are equal. In the tests discussed here, the null hypothesis that the variances are equal could not be rejected

Creating a Global Flow Variable

One approach to using flow variables is to make the variable available to all nodes in a workflow. To create one or more global flow variables in a workflow, open the workflow by right-clicking it in the Explorer panel. Then follow these steps.

• In the KNIME Explorer panel, right-click on the workflow in which the flow variable is needed. This opens the Workflow Variable Administration panel.
• Click on the Add button.
• Enter the desired name for the global flow variable in the Variable namespace.
• Select the Variable Type using the drop-down selector (DOUBLE, INTEGER, or STRING).
• Enter the default value for the global flow variable.
• Click OK, then OK again.

Creating a String Flow Variable

The following workflow (Fig. 3.15) uses a String Configuration node so that a user can select state (IL, IN, or OH) from an insurance file and then create two Excel files, one for the continuous variables and one for the nominal (string) variables. Descriptions of the nodes in the workflow are in Table 3.5.

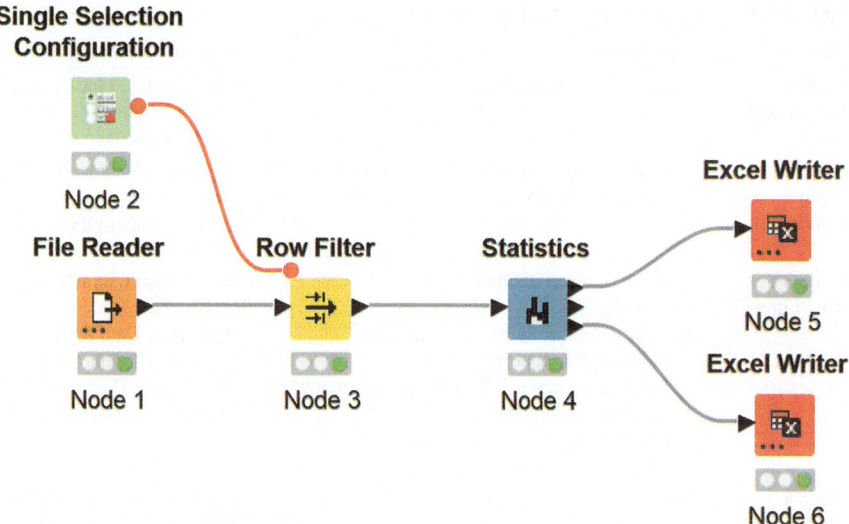

Fig. 3.15 Using a string flow variable

A single worksheet is created in each Excel file with the state's name where the policy was written. This demonstrates the use of a flow variable to enable the selection of specified rows. The flow variable is passed to the Row Filter to select the state that will be forwarded to the rest of the workflow. The advantage of using a flow variable in this example is that the choice is made in one place and can be used downstream of the workflow.

3.9   Loops in KNIME

In some workflows, there is a need to repeat the execution of a set of nodes until a certain condition is met. The general structure of a loop is shown in Fig. 3.16.

A loop has a starting point, one or more processing nodes, and an end point. Several different conditions can be set in KNIME as shown in Table 3.6.

Run a Loop a Preset Number of Times
The Counting Loop executes the workflow between the Start and End a specified number of times. In the example shown in the workflow in Fig. 3.17, the result of 2 to the power of 0 through 9 is found, and the results are collected in the output of the Loop End. Node descriptions are in Table 3.7.

Table 3.5 Node descriptions for the Fig. 3.15 workflow

Node	Descriptions	Input and Output ports
Node 1 File Reader	Read file: FraudData.csv	Output port: FraudData data table with 1000 rows and 38 variables
Node 2 String Configuration	Output a string flow variable with a set value.The value can be IL, IN, or OH	Input port: Data table from the Node 1 Flow variable output port: policy_state
Node 3 Row Filter	Filter rows using defined criteria Filter Criteria: Select: Include rows by attribute value Column to test: policy_state Matching criteria: Select use pattern matching Mouse-click on V icon Check: Use Variable Select: string-input	Flow variable input port: policy-state Input port: Data table from the Node 1 Output port Output port: Data table with Filtered rows
Node 4 Statistics	Options Include: age policy_state insured_sex insured_education_level total_claim_amount Exclude all other variables	Input port: Data table from the Node 3 Output port Upper output port: Continuous statistics Middle output port: Not used Lower output port: Occurrences
Node 5 Excel Writer	The Excel file will be stored with the workflow for future use. The results are written to separate sheets in the Excel file: Statistics and Occurrences Settings: Excel format: XLSX Write to Relative to Current workflow data area File: Statistics.xlsx sheet name: Statistics Write options: If exists: Overwrite Flow Variables Sheet_names: string-input (in a grey area)	Input port: Descriptive statistics from Upper output port of Node 4
Node 6 Excel Writer	The Excel files will be stored with the workflow for future use Settings Excel format: XLSX Write to Relative to Current workflow data area File: Statistics.xlsx sheet name: Occurences Write options: If exists: Overwrite Flow Variables Sheet_names: string-input (in the grey area)	Input port: Descriptive statistics from Lower output port of Node 4

Fig. 3.16 Structure of
KNIME loops

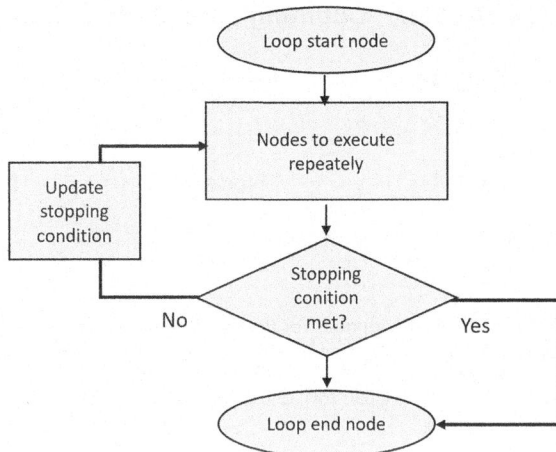

Table 3.6 Examples of KNIME loop structures

Type of loop	Loop start node	Loop end node
Run a loop a preset number of times	The Counting Loop Start This node starts a loop which is executed a predefined number of times specified in the Counting Loop Start dialog	Loop End This node marks the end of a workflow loop and collects the intermediate results by row-wise concatenation of the incoming tables
Loop on a set of columns	Column List Loop Start This node iterates over a list of columns of the input table. Its columns are divided into two groups, the "excluded" columns will always be included in any iteration and the "included" columns will be visible once in each iteration	LoopEnd (Column Append) This node collects the intermediate results by joining the tables based on their row IDs. In each iteration, the node will join the current input table with the previous result
Loop until a condition is satisfied	Generic Loop Start This creates a loop that is executed until a specified condition on one of the flow variables is met. The condition must be set in the loop's end node	Variable Condition Loop End This node executes the loop's body until a certain condition on one of the flow variables is met

Loop on a Set of Columns

Running a set of columns (variables) through the same workflow is sometimes convenient. This can be done with the Column List Loop. In the simple demonstration in Fig. 3.18, five columns of random numbers, normalized to a range of 0–1.0 are created. The values are then run through a loop that multiplies each number by 1000 and rounds the result to two decimal places using a Math Formula node. The

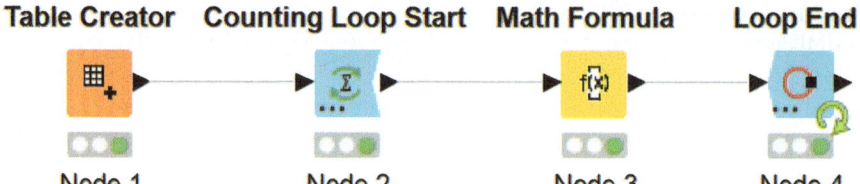

Table Creator   Counting Loop Start   Math Formula   Loop End

Node 1   Node 2   Node 3   Node 4

Fig. 3.17 A simple counting loop

Table 3.7 Node descriptions for the Fig. 3.17 workflow

Node	Descriptions	Input and output ports
Node 1 Table Creator	Create a one-row, one-column starting value of 1.0	Output port: Data table with 1 row and 1 column
Node 2 Counting Loop Start	Starts a loop and runs it a set number of times Standard settings: Number of loops: 10	Input port: Data table from the Node 1 Output port Output port: Data table plus a flow variable with the current iteration number and the maximum number of iterations
Node 3 Math Formula	Create a math expression; note that the Math Formula node automatically inputs and outputs flow variables Flow Variable List: maxIterations currentIteration Column List: $2^{\text{power of iteration}}$ Expression : 2 ^ $\$\$\{currentIteration\}\$\$$ Select: Replace Column with column name $2^{\text{power of iteration}}$ Output port: $2^{\text{power of iteration}}$ Flow variables output maxIterations currentIteration	Input port: Data table from the Node 2 Output port Output port: $2^{\text{power of iteration}}$
Node 4 Loop End	Mark the end of the loop and collect the results Standard settings: Use row IDs by appending a suffix Check Add iteration column Check Ignore empty input tables	Input port: Data table from the Node 3 Output port Output port: Table with $2^{\text{power for iterations}}$ for iterations = 0–9

workflow is shown in Fig. 3.18, and node descriptions are in Table 3.8. The data generator output is in Table 3.9, and the resultant output of the loop is in Table 3.10.

Loop Until a Condition Is Satisfied

This loop is executed until a preset condition is reached on one of the flow variables. The condition can be set to stop when the selected variable is $<$, $>$, $=$, $\leq$, or $\geq$ a

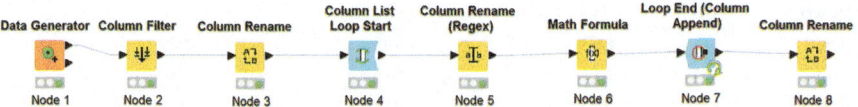

Fig. 3.18 Example of Column List loop

Table 3.8 Node descriptions for the Fig. 3.18 workflow

Node	Descriptions	Input and output ports
Node 1 Table Creator	Creates three variables with four observations, each with a range of 0–1	Output port: Data table with 3 columns and 4 rows
Node 2 Column List Loop Start	Starts a loop that iterates over a list of columns Standard settings: Include: Var1 to Var3 Exclude: No entry	Input port: Data table from the Node 1 Output port Output port: Data table with a column for each iteration
Node 3 Column Rename	Rename each input variable Var1 to Var3 temporarily to new In Flow variable dialog change old_column_name to currentColumnName	Input port: Data table from the Node 2 Output port Output port: Data table with one column labeled new
Node 4 Math Formula	Math expression: Expression: round(1000*new,0) Replace Column: new	Input port: Data table from the Node 3 Output port Output port: Data table with variable new × 1000 rounded
Node 5 Column Rename	Rename variable new to Var1 to Var3 In Flow variable dialog, change new_column_name to currentColumnName	Input port: Data table from the Node 4 Output port Output port: Data table with renamed column
Node 7 Loop End (Column Append)	Ends the Column List Loop Loop End Configuration Use defaults	Input port: Data table from the Node 5 Output port Output port: Collected results

Table 3.9 Input values to the loop in workflow 3.8

RowID	V1	V2	V3
Row0	0.744	0.931	0.319
Row1	0.663	0.995	0.323
Row2	0.611	0.846	0.154
Row3	0.729	0.915	0.252

constant value. Figure 3.19 shows a simple example of a conditional loop, and Table 3.11 has the node descriptions.

The loop was run 27 times until a value of 0.083 was generated, stopping the process. (Different results will occur with future runs because of the random value for the seed.)

Table 3.10 Output values of the loop in workflow 3.8

RowID	V1	V2	V3
Row0	744	931	319
Row1	663	995	323
Row2	611	846	154
Row3	729	915	252

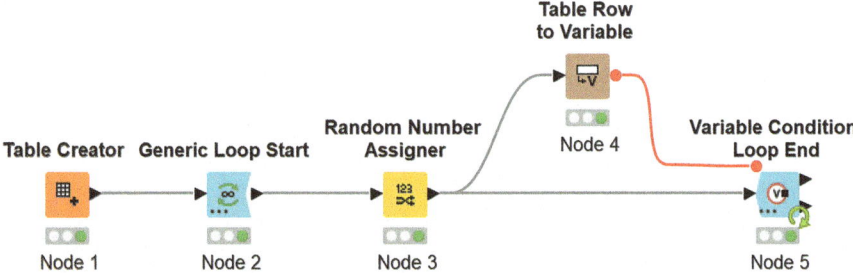

Fig. 3.19 Example of a loop until the condition is met

Table 3.11 Node descriptions for the Fig. 3.19 workflow

Node	Descriptions	Input and output ports
Node 1 Table Creator	Create a one-row, one-column starting value of 1.0 with name Var 1	Output port: 1 row by 1 column data table
Node 2 Generic Loop Start	This starts the looping process. The data from the Table Creator is just passed through. No settings are needed	Input port: Data table from the Node 1 Output port Output port: Data table from the Node 1 Output port
Node 3 Random Number Assigner	Assign a uniform random value Options: Column Name: newvalue Seed: Do not check so that a new seed will be selected with each loop iteration Name: for all Min: 0.0 Max: 1.0	Input port: Data table from the Node 2 Output port Output port: Table with a random variable 1 row by 2 columns: Var 1 and newvalue
Node 4 Table Row to Variable	Convert the first row of a data table to a flow variable Settings Include: newvalue Exclude: Var 1 All other settings to default values	Input port: Data table from the Node 3 Output port Flow variable output: Random variable newvalue 1 row by 1 column
Node 5 Variable Condition Loop End	This is where the loop parameters are set Default settings: Available flow variables: newalue Finish loop if the selected variable is: <0.1 Select: Collect rows from the last iteration	Input port: Data table from the Node 3 Output port Flow variable input: Data table from Flow variable output of Node 4 Upper output port: Collected results Lower output port: Collected results with Iteration number

3.10   Metanodes and Components in KNIME

The workflow shown in the previous examples demonstrated how KNIME was used to construct analytics models as a series of connected nodes. The nodes are placed using a drag-and-drop process resulting in a self-documented workflow diagram. While this feature is a clear advantage of using KNIME, the downside is that the workflows can become complex, sometimes to the point where they are confusing. To address this problem, KNIME includes two tools to clean up messy workflows, Metanodes and Components, both of which can bundle several nodes into a single node. A Metanode simply groups multiple nodes or even other Metanodes, while Components can encapsulate more complex processes.

Metanodes
Metanodes have a distinct appearance in KNIME and have three states: idle, executing, and executed. The icons for each state are shown in Fig. 3.20.

There are many possible configurations of Metanodes, with differing numbers and types of input and output ports. Four basic structures are shown in Fig. 3.21.

Table 3.12 shows the steps needed to create a Metanode from three nodes using the Fig. 3.10 workflow.

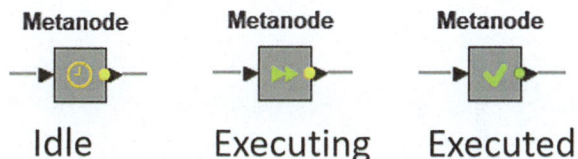

Fig. 3.20 Three states of Metanodes

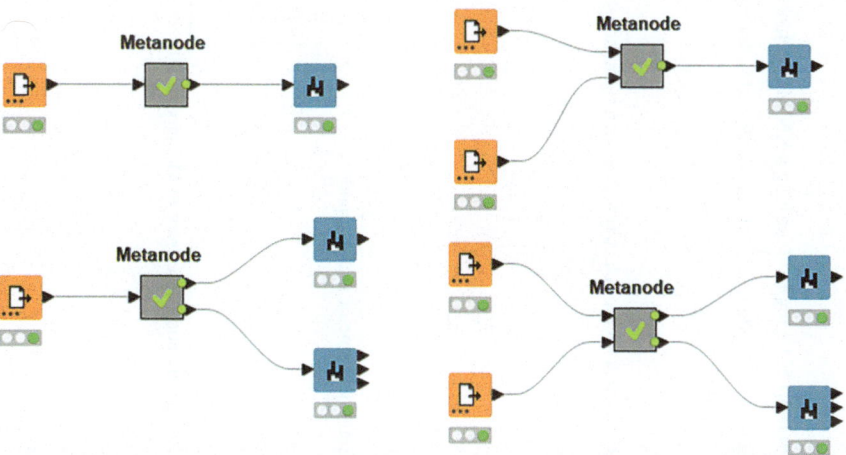

Fig. 3.21 Four Metanode structures

Table 3.12 Example of creating a KNIME Metanode

Select each node for the Metanode by right clicking the mouse while holding down the Ctrl key	
Right-click on the selected nodes, then right mouse click on create Metanode	
Enter the name of the Metanode	
Change the node descriptor to Metanode	

The Metanode in Table 3.12 replaces the three String Manipulation nodes from Fig. 3.10, resulting in the same variable transformations.

Components

Like Metanodes, Components can also group nodes. But components can also include configuration dialogs which make the workflow interactive. The following example uses the insurance fraud data discussed previously to demonstrate the using a Component with a list box configuration selector.

The workflow in Fig. 3.22 reads FraudData.csv that was saved in the current workflow area. Node descriptions are in Table 3.13. The Variable Selection Configuration provides a drop-down menu to select a column from the variable list and then a specific value from the column. The column and value selections are sent to the Row Filter to create a subset of the data. The output of the Row Filter is sent to the Statistics node. For example, using the Fraud Data set, the column variable might be "make" and the selected value "Chevrolet." Then, the Statistics node calculates descriptive statistics on all continuous variables and frequencies on all categorical variables. The upper Output port of the Statistics node has the table with statistics on continuous variables, and the lower Output port contains the frequencies for the categorical variables.

The column and value selections flow variables are sent to the Variable to Table Column nodes (one each for the statistics output of the Statistics node and one for the Occurrences output). The Variable to Table Column nodes add the column and selection values to the data table created by the Statistics node. Then an Excel file is created with two sheets, one for the statistics output and one for the occurrences.

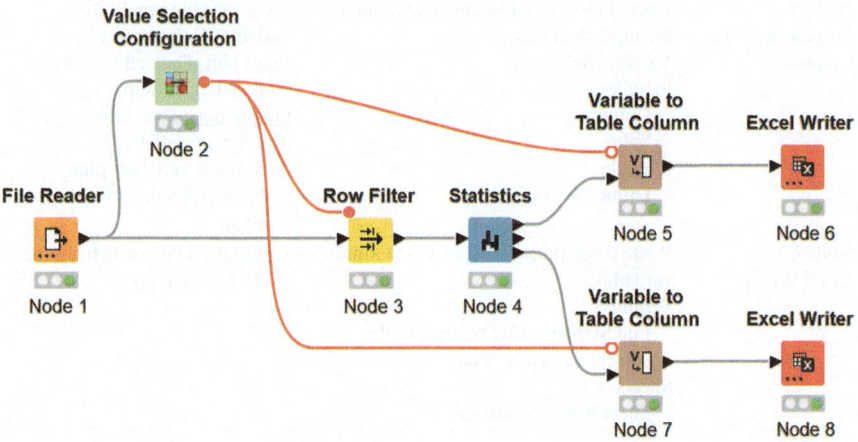

Fig. 3.22 Example of a workflow without using a component

Table 3.13 Node descriptions for the Fig. 3.22 workflow

Node	Description	Input and output ports
Node 1 Table Creator	Read FraudData.csv	<u>Output port:</u> Data table with 1000 rows and 38 variables

Node 2
Value Selection
Configuration
	Provide a column selection configuration for use in a component. Outputs a string flow variable with the name of the selected value
	Control:
	  Label: ColumnSelection
	  Description: Select column and value to subset FraudData.csv table
	  Parameter/Variable Name: select
	  Selection Type: Drop-down
	  Default column: policy_state
		<u>Input port:</u> Data table from the Node 1 Output port
		<u>Flow variable output:</u> Flow variables Column and Value

Node 3 Row Filter	This node subsets the FraudData table using the Column variable and its value (policy_state) Flow variable input: Column and Value selections from Node 2	<u>Input port:</u> Data table from the Node 1 Output port <u>Output port:</u> Subset of FraudData table for the selected state
Node 4 Statistics	Computes statistics on continuous and categorical variables Options: Include: All variables	<u>Input port:</u> Data table from the Node 3 Output port <u>Upper output port:</u> Table with descriptive statistics on continuous variables <u>Middle output port:</u> Not used <u>Lower output port:</u> Table with occurrences (frequencies) for all variables

Node 5
Variable to Table
Column
	Extract flow variables and add them to the input data table
	Variable Selection:
	Include:
	  Column
	  Value
	Exclude
	  knime_workspace
		<u>Flow variable input:</u> Flow variables Column and Value
		<u>Input port:</u> Data table from the Node 4 Upper output port
		<u>Output port:</u> Data table with descriptive statistics on continuous variables plus Column and Value in last two columns

Node 6
Excel Writer
	Write descriptive statistics on continuous variables
	Settings:
	  File Statistics.xlsx relative to the current workflow area
	Sheets:
	  Sheet name: Statistics
		<u>Input port:</u> Data table from the Node 5 Output port

(continued)

Table 3.13   (continued)

Node	Description	Input and output ports

Node 7
Variable to Table
Column
	Extract flow variables and add them to the input data table. This is done so that the Excel worksheets contain the selected column and value
	Variable Selection
	Include:
	   Column
	   Value
	Exclude:
	   knime_workspace
		Flow variable input: Flow variables Column and Value
		Input port: Data table from the Node 4 Lower output port
		Output port: Table with occurrences on all variables

Node 8
Excel Writer
	Write frequency statistics in an Excel file
	Settings
	   File Statistics.xlsx relative to the current workflow area
	Sheets:
	   Sheet name: Occurences
		Input port: Data table from the Node 7 Output port

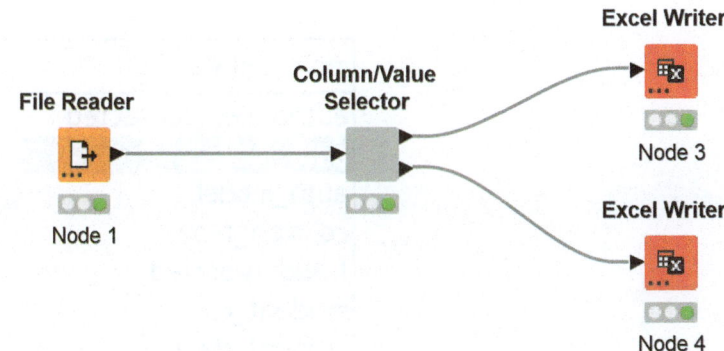

Fig. 3.23   Workflow with component replacing nodes in Fig. 3.22

Creating a Component for the nodes in Fig. 3.23 involves the following steps:

- Select the nodes 2, 3, 4, 5, and 7 for the Component by right clicking the mouse while holding down the Shift key and then selecting Create Component.
- After entering a name (Column/Value Selector) in the dialog box, click OK, and the Component will appear in the workflow, shown in Fig. 3.23.

To use the Component in the workflow, right-click on it and select Configuration. The following Options will be shown (Fig. 3.24):

Two drop-down lists appear the upper one for the column (variable) and the lower one to select the specific value for the column. In this example, "auto-make" is selected Fig. 3.25.

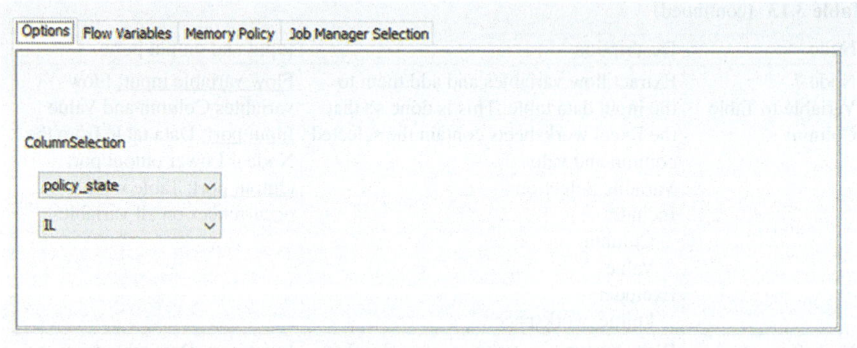

| Options | Flow Variables | Memory Policy | Job Manager Selection |

ColumnSelection

policy_state ∨

IL ∨

Fig. 3.24 Options after clicking on Column/Value Selector

Fig. 3.25 Drop-down list
to select an auto-make

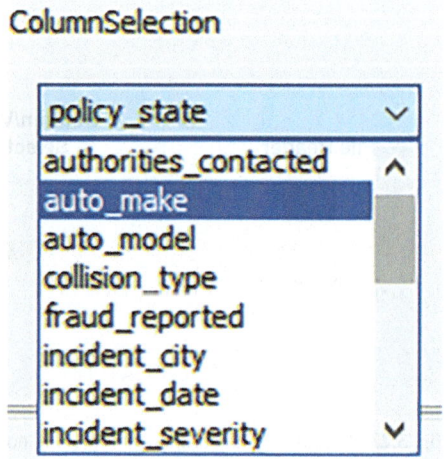

After selecting "auto-make," a second drop-down list appears with the options available for that column. "Chevrolet" was selected for this example Fig. 3.26.

Next, Right click on the Column/Value Selector and select Execute. To create the Excel files, right-click on Node 3 to create an Excel file for the continuous variables and on Node 4 to create the Excel file for the occurrences table.

The component-based workflow in Fig. 3.23 contains the same processing steps as the original workflow. To look inside the Component, right-click on it and select Component =>Open. If you click Component =>Expand instead, this will restore the workflow nodes shown in Figure (except that some repositioning of the nodes may be needed).

Fig. 3.26 Drop-down list
to select an auto brand

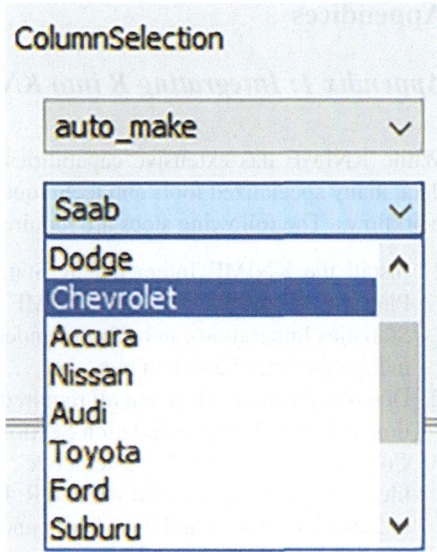

3.11   Summary

KNIME will be used extensively throughout this text for the analyses. Key strengths
of KNIME are the graphical workflow interface, open-source extensibility, ease of
use, and breadth of analytics methods available. An extensive array of free resources
such as courses, documentation, and a community forum support learning and using
KNIME. Several features of KNIME were demonstrated using simple examples,
but a more in-depth understanding can be gained by viewing the online tutorials
from the KNIME site. At the least, it is important to download and install KNIME
before continuing with the following chapters and becoming familiar with KNIME
by working on some problems.

   In this chapter, links to obtain and install KNIME were provided as well as links
to a variety of tutorials and documents available to learn how to use the software.
Features of KNIME were demonstrated, including using loops to repeat processes,
Metanodes to organize workflows, and Components to provide interactive configuration dialogs.

Appendices

Appendix 1: Integrating R into KNIME

While KNIME has extensive capabilities for preparing, viewing, and analyzing data, many specialized tools and techniques from R can be integrated into KNIME workflows. The following steps are required to incorporate R into KNIME.

1. Install the KNIME Interactive R Statistics Integration. In KNIME Analytics Platform, go to File → Install KNIME Extensions…The KNIME Interactive R Statistics Integration can be found under KNIME & Extensions or by entering R integration into the search box.
2. Download and install R and all required packages as described in the R installation and R packages installation Sections.
3. Configure the KNIME Interactive R Statistics Integration by selecting file => Preferences => KNIME and R. Enter the path to R and set the buffer size to 256. Click Apply and then Apply and Close (Fig. 3.27).

At this point, the base functions in R will be available in KNIME. Thousands of optional R packages can also be used in KNIME. Any optional package must be installed using R and n loaded using the R library() function. These will be indicated in the text where optional packages have been used.

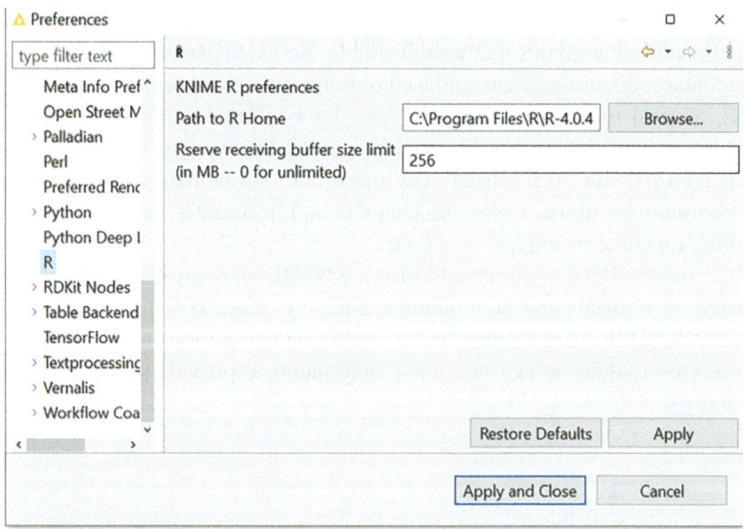

Fig. 3.27 Configuring the R extension

Appendix 2: Regex for Search Patterns

A *Regular Expression* is a pattern to specify a certain text string and a way to find, format, and manipulate text. Regular Expressions (abbreviated as regex) are used in several KNIME nodes, including Column Filter, Row Filter, Column Rename (Regex), and Constant Value Column Filter, to perform transformations and searches. Regex is somewhat confusing at first, but it is useful to learn to take full advantage of KNIME. Luckily, several free online tools are available to test your regex before using it in a KNIME node. A good site to use is Regular Expressions 101.[14] Also recommended is an excellent introduction from O'Reilly: "An introduction to regular expressions."[15]

Regex is used in many programming languages in addition to KNIME. Unfortunately, the syntax differs for Python, C#, Ruby, PHP, and other implementations. A partial list of Regex syntax for KNIME is shown in the following Table 7.14. Examples of using the syntax are shown in Table 7.15.

Table 7.14 Regex Syntax

Syntax	Description
?	Optional part of expression; match 0 or 1 time
*	Match 0 or more times
+	Match 1 or more times
[ ]	Range of acceptable values
{ }	Repeat with exactly n characters
		Alternative execution (or)
( )	Grouping characters or extracting substrings
i	Case insensitive
^	Anchor to the beginning of the string or negation
$	Anchor to the end of the string
.	Arbitrary character

---

[14] https://regex101.com/

[15] https://www.oreilly.com/content/an-introduction-to-regular-expressions/

Regex Examples

Table 7.15 Examples of regex using KNIME node Column Rename (Regex)

Match	Syntax	Replace with	Old name	New name
letter **e** at end of name	e$	e2	Table	Table2
			Trade	Trade2
			Test	Test
			Toon	Toon
			Talent	Talent
match a through g	[a-g]	-	Table	T—l-
			Trade	Tr----
			Test	T-st
			Toon	Toon
			Talent	T-l-nt
match o	[o]	e	Table	Table
			Trade	Trade
			Test	Test
			Toon	Teon
			Talent	Talent
match o twice	[o]{2}	e	Table	Table
			Trade	Trade
			Test	Test
			Toon	Ten
			Talent	Talent
Match T	[T]	t	Table	table
			Trade	trade
			Test	test
			Toon	toon
			Talent	talent
match t at the end of a string	[t]$	T	Table	Table
			Trade	Trade
			Test	TesT
			Toon	Toon
			Talent	TalenT
remove spaces from name replace with blank	\s		Table	Table
			Trade	Trade
			Test	Test
			T o n	Toon
			T alent	Talent

Chapter 4
Data Preparation

4.1 Obtaining the Needed Data

One aspect of the analytics process that may be surprising is that a company's or organization's data warehouse may not provide the needed data in the correct form. The data must be assembled, cleaned, and custom-fitted to the analytics problem. This process is most efficient and effective when the analyst knows the business well or if a domain expert is on the project team.

In addition, external data from the web, government, and commercial sources may be needed and integrated into a functional structure with data from an organization's internal systems. External weather, traffic, calendar, or census data might be required. Functions such as joins can link records between two or more databases by matching key fields.

Most predictive and data mining software, including KNIME, requires the data to be analyzed in a well-formed tabular structure or flat file. Variables should be in the columns and observations in the rows. The data table must be transposed if the data source has variables in the rows and observations in the columns.

It is also imperative that the proper type of data is available. For example, if a predictive model is to be developed, it is necessary to have both predictors and outcome variables. If a binary outcome is to be predicted, both binary values must have enough instances. For example, a predictive model to identify fraud requires both fraudulent and not- fraudulent observations.

As the data is assembled, it is sometimes discovered that the available data is not appropriate to address the defined problem. In this case, the business problem might have to be revised or new ways explored to obtain the required data.

© The Author(s), under exclusive license to Springer Nature Switzerland AG 2023   53
F. Acito, *Predictive Analytics with KNIME*,
https://doi.org/10.1007/978-3-031-45630-5_4

4.2 Data Cleaning

A data set might include inaccurate, incomplete, or inconsistent values, duplicate records, spelling errors in texts or labels, or simply errors from the original data entry. Analyzing raw data with such mistakes can lead to incorrect or misleading results. Data cleaning (aka data scrubbing or data cleansing) is important before a model is built. It is not as simple as deleting records with apparent problems. Instead, it is important to maximize accuracy while not removing valuable information. The objective is to ensure that the data set used for modeling is of the highest quality. Some of the activities conducted in the data cleaning process include:[1]

- Removing unneeded columns (variables) and or rows.
- Removing duplicate observations.
- Removing variables that have constant values for all observations.
- Removing data values that are out-of-scope and not within the required time frame.
- Finding and correcting impossible or non-sensible values where logical constraints can be specified. (For example, age in years should equal current year minus birth year, population density should equal total population divided by area, etc.)
- Identifying and dealing with outliers.
- Identifying and dealing with missing values.
- Finding misspellings, out-of-range, or impossible values. (For example, ensuring that dates, geographic regions, and other variables fall into acceptable ranges.)
- Making sure that the formats and units of measurement are consistent. (For example, have the variables measured the same way over time, or have the data definitions been changed? Or are the variables from different data sets available at the same level of granularity?)
- Ensuring that regular expressions such as phone numbers, zip codes, social security numbers, etc., are in a consistent, correct, and usable format.

The preceding list of things to check may seem imposing, but it is only a partial list of necessary steps. Also, some of the listed activities may not be relevant to a particular project, so there is no one-size-fits-all process to follow since it depends on the specific problem definition and predictive models used.

Sage Advice

Give me six hours to chop down a tree and I will spend the first four sharpening the axe. (Abraham Lincoln)

4.3 Data Cleaning Nodes in KNIME

KNIME provides several nodes that can be used for cleaning the data in Fig. 4.1.

---

[1] Source of quote: https://www.setquotes.com/give-me-six-hours-to-chop-down-a-tree/

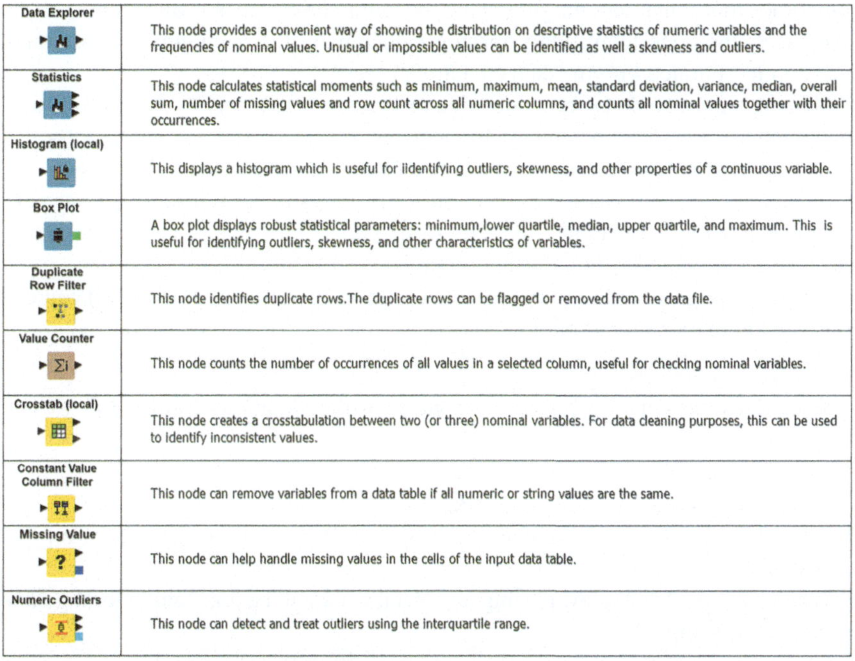

Data Explorer	This node provides a convenient way of showing the distribution on descriptive statistics of numeric variables and the frequencies of nominal values. Unusual or impossible values can be identified as well a skewness and outliers.
Statistics	This node calculates statistical moments such as minimum, maximum, mean, standard deviation, variance, median, overall sum, number of missing values and row count across all numeric columns, and counts all nominal values together with their occurrences.
Histogram (local)	This displays a histogram which is useful for lidentifying outliers, skewness, and other properties of a continuous variable.
Box Plot	A box plot displays robust statistical parameters: minimum,lower quartile, median, upper quartile, and maximum. This is useful for identifying outliers, skewness, and other characteristics of variables.
Duplicate Row Filter	This node identifies duplicate rows.The duplicate rows can be flagged or removed from the data file.
Value Counter	This node counts the number of occurrences of all values in a selected column, useful for checking nominal variables.
Crosstab (local)	This node creates a crosstabulation between two (or three) nominal variables. For data cleaning purposes, this can be used to identify inconsistent values.
Constant Value Column Filter	This node can remove variables from a data table if all numeric or string values are the same.
Missing Value	This node can help handle missing values in the cells of the input data table.
Numeric Outliers	This node can detect and treat outliers using the interquartile range.

Fig. 4.1 Selected KNIME nodes useful for data cleaning

4.4   Missing Values

The presence of missing values is one of the most challenging problems in analytics. In an ideal situation, there would be valid values for every observation. However, some missing values will be present for some of the variables in most real-world data sets.

If missing values are found, the best thing to do is go back to sources and find the value for the missing information. While it would be great if the required values could be found, in practice, it is unlikely.

Missing values can be indicated in many ways in a data set, so analysts need to look for such indicators. Some of the ways that missing values are indicated in data sets include the following:

Null
99, 999, −999, etc.
−1
?
●
NA
None
NaN

Sometimes more than one missing value indicator can occur in a single data set. In such cases, the code identifying missing values may differ among variables, and must be incorporated into the data cleaning process.

Missing values could be generated due to structural reasons. For example, a variable might ask whether a person is a homeowner. If yes, a question might ask for the home's approximate value. If not, the home's value is missing, as it should be.

In survey research, it is quite common for respondents to refuse to answer personal questions such as annual income. Respondents may not know the answer to the question or consider the information private.

When missing values are found in a data set, consideration of the following questions can aid in data preparation:

- Why is the data missing?
- What does "missing" mean in the context of the situation? Refusal? Does not apply? Error? Not collected?
- How many cases contain missing data?
- Are values missing for key variables?
- What if cases with missing data are dropped? What will be the impact on the analysis?
- When the model is deployed using new data, what will be done with missing data in the new sample?

Types of Missing Values

A classification scheme for missing values developed by Rubin (1976) stipulated three types of missing values:

- MCAR: Missing completely at random.
- MAR: Missing at random.
- MNAR: Missing, not at random.

MCAR is the most favorable situation since the missing values are not related to the target or predictor values. The observations with missing values should have similar distributions as those without since the missing values in the rows of a data set are assumed to result from random processes. For example, if a sensor is used to collect measurements in a manufacturing process, there may be random malfunctions of the sensor itself. Despite the convenience of assuming MCAR, however, it is often unrealistic.

MAR means systematic differences exist between the observations with and those without missing values. MAR assumes that missing values in a variable are related to values of other predictor variables. For example, income might be related to education and age. An estimate of income could be derived from known values of age and education. MAR is a more realistic assumption than MCAR, and many missing data imputation methods start with the MAR assumption.

MNAR refers to situations where the likelihood that a value is missing is not related to other variables but to the variable's value. For example, in a data set on creditworthiness, a question on past personal bankruptcies may be left blank by persons who did declare bankruptcy. MNAR is the most pernicious type of missingness. It is difficult to determine whether missing values are MNAR versus MAR or MCAR, but domain knowledge can help identify when MNAR is present for a specific variable.

One approach to distinguish between Missing Completely at Random versus Missing at Random was suggested by Little and Rubin (1989). Their approach is to split the data set into two groups: one group with the variable in question not missing and the other with the variable missing. Then, tests are run between the groups on different characteristics that are present in the data. If significant differences are found, then the data is not Missing Completely at Random but likely Missing at random.

Handing Missing Values
If the proportion of observations with missing information is small in a large data set (e.g., less than 1%), those observations may sometimes be deleted. Deleting such observations is known as *listwise deletion*. Before dropping cases, however, it is essential to check whether the missing values are concentrated in just a couple of variables or if many or most of the observations have at least one missing value. Unfortunately, in the latter case, it is possible that a data set with thousands of observations could be reduced to only hundreds using listwise deletion, even with a small percentage of missing values,. For example, assume that each of the 40 variables in a sample of 10,000 cases has 5% of the values missing spread randomly and independently throughout. The chance that a given record will have complete data and thus escape being deleted is $(0.9540)^{40} = .1285$. Therefore, more than 85% of the cases may be dropped, leaving only a sample of 1285 cases.

Another approach to consider is to delete columns with large percentages of missing data. For example, if 75% of observations on a particular variable are missing, the column might be deleted. This, of course, should not be done if the situation is deemed MNAR. The presence of a missing value can itself be important and predictive.[2]

Finally, there are methods for imputing values for missing data. Imputation assumes values are missing at random. The advantage of imputation is that observations are not dropped from the analysis, which could lead to bias or poorer estimates. However, imputation can reduce variability and decrease correlations with other variables.

Excessive Missing Values
In some cases, columns (variables) can have a large percentage of missing values. For example, imputing the missing information might distort the analysis results if the percentage of missing values in a variable is greater than 50%, although, to be clear, this is an arbitrary threshold. One approach is to compare the results of an analysis with the variable dropped with the results if the missing values are imputed.

---

[2] It is possible to empirically evaluate what threshold level missing values would be optimal by running a model with varying the percentage from 0% to 100% and checking model performance.

Table 4.1 Data table created for missing value demonstration

RowID	Var1	Var2	Var3	Var4	Var5	Proportion of row missing
Row0	10	a	a	1	3.6	0.0
Row1	6	b	b	?	8.3	0.2
Row2	?	c	?	?	2.1	0.6
Row3	4	?	?	3	?	0.6
Row4	11	e	d	?	5.3	0.2
Row5	?	?	d	?	?	0.8
Row6	9	g	g	?	6.5	0.2
Row7	12	a	d	?	7.3	0.2
Row8	?	s	?	6	3.4	0.4
Proportion of column missing	0.33	0.22	0.33	0.67	0.22	

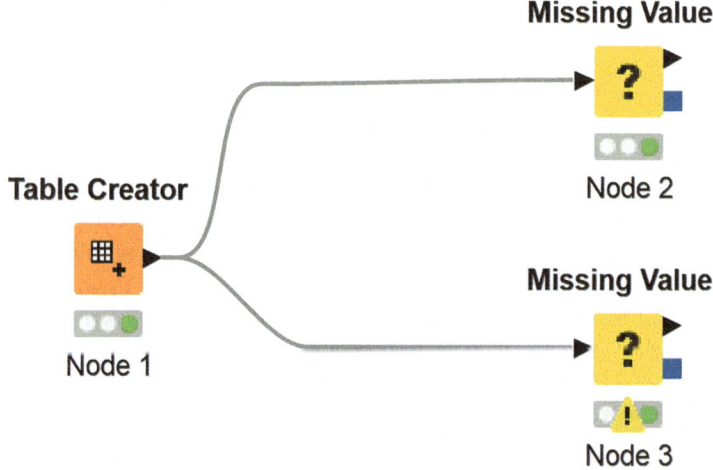

Fig. 4.2 Using the Missing Value node

The following data table (Table 4.1) with several data types and arbitrary numbers of missing values was created using the Table Creator node to demonstrate some of the methods available in KNIME. Question marks indicate the missing values in the data table.

The workflow in Fig. 4.2 shows two uses of the Missing Value node. In the upper workflow, imputations were made for the integer, double, and string variables. In the lower workflow, listwise deletion was used to delete any row with one or missing values. The node descriptions are in Table 4.2.

The cell values imputed from the upper output port of the Fig. 4.2 workflow are shown in bold in Table 4.3. Notice that row 5 now has no missing values, but all but one value was imputed. Variable Var4 had only three values, the median of which was 3. Now all but two values are threes. While this demonstration shows how imputation works with the Missing Value node, it also shows that the results may not be desirable.

Table 4.2 Node descriptions for the Fig. 4.2 workflow

Nodes	Descriptions	Input and output ports
Node 1 Table Creator	Create five variables: Var1-Var5 with missing values	Output port: Data table with created values (9 rows by 5 columns)

Node 2 Missing value	Default: imputation:
	    Number (double): Mean
	    String: Most frequent value
	    Number (integer) Median
		Input port: Data table from the Output port of Node 1
		Output port: Table with imputed values for missing data (9 rows by 5 columns)

Node 3 Missing Value	Default: Remove rows with missing values as follows:
	    Number (double): Remove Row
	    String: Remove Row
	    Number (integer) Remove Row
	Note that an error is raised here. That signals that the value imputation methods being used will not work in PMML 4.2 the standard for exchanging models among different programs. This is not a problem in the present example
		Input port: Data table from the Output port of Node 1
		Output port: Table with rows removed that had missing values (1 row by 5 columns)

Table 4.3 Imputation results

RowID	Var1	Var2	Var3	Var4	Var5
Row0	10.0	a	a	1	3.60
Row1	6.0	b	b	3	8.30
Row2	**8.7**	**c**	**d**	3	2.10
Row3	40.0	a	d	3	5.21
Row4	11.0	e	d	3	5.30
Row5	**8.7**	a	**f**	**3**	5.21
Row6	9.0	g	d	3	6.50
Row7	12.0	a	c	3	7.30
Row8	**8.7**	**s**	**d**	6	3.40

Table 4.4 Results with row removal for at least one missing value

RowID	Var1	Var2	Var3	Var4	Var5
Row0	10	a	a	1	3.6

Table 4.4 shows the data table resulting from the lower output port of Fig. 4.2 after any row with at least a single missing value was removed. Only Row0 was retained, again demonstrating that care must be taken using the Missing Value node.

Dropping Rows with Greater than a Specified Proportion of Missing Values
Instead of dropping rows with only one or more missing values, it might be desirable to drop rows only if the proportion of missing values is greater than some pre-specified rate. Another approach might be to drop columns only if the proportion of

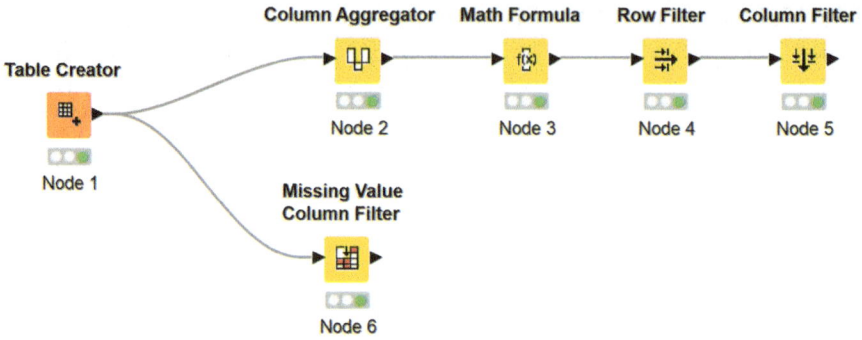

Fig. 4.3 Workflows to drop rows or columns with an excessive number of missing values

missing values in the column is greater than a given rate. The upper workflow in Fig. 4.3 drops rows if the proportion of missing values in a row exceeds a threshold set in the Row Filter. The lower workflow in Fig. 4.3 removes columns where the proportion of missing values exceeds a threshold. Node descriptions are in Table 4.5.

Table 4.6 shows the output of Node 5 in the Fig. 4.3 workflow. A column with the proportion of missing values was added to each row. Table 4.7 shows the two columns with less than 30% missing values.

Flagging Missing Values

The fact that a column has a missing value might be informative, so capturing that information could be helpful. For example, the variable "income" might be missing in a loan default prediction model. The fact that the variable is missing can signal a problem and predict a likely default. The workflow in Fig. 4.4 shows how to convert a variable with a high percentage of missing values into a string variable that indicates "Missing" or "Not missing." The column with the high percentage of missing values is converted to a string variable using Number To String. Then the Rule Engine is used to create the binary string indicating whether the entry is missing. Node descriptions are in Table 4.8, and the output from Node 3 is in Table 4.9.

Var4 Missing can now be used as an indicator in place of **Var4**. If **Var4 Missing** is used in predictive models, then **Var4** should not be included.

4.5 Dealing with Missing Values

The KNIME node "Missing Value" handles missing values for numeric and string variables. The node can replace all numeric and string variables in the same manner. Or the type of imputation can be set differently for each variable.

Imputation refers to substituting a value to replace the missing values in a data set. (If MNAR is suspected, the imputation techniques discussed in the following should not be used since the results of analyses can be biased and misleading.)

Table 4.5 Node descriptions for the Fig. 4.3 workflow

Nodes	Descriptions	Input and output ports
Node 1 Table Creator	Create five variables with Var1–Var5 having missing values	Output port: Data table with created values (9 rows by 5 columns)

Node 2
Column Aggregator
	Groups and aggregates the columns listed
	Columns
	  Aggregation columns
	  Var1–Var5
	Options:
	  Column name: Count
	  Aggregation method:
	  Count
	  Check Missing box
	  Column name: Missing
	  value count
	  Aggregation method:
	  Missing value count
		Input port: Data table from the Output port of Node 1
		Output port: Input table plus Counts of values and Missing values.

Node 3
Math Formula
	Calculate the proportion missing
	Math Expression
	  Table with created values
	  Append Column:
	  Proportion missing
		Input port: Data table from the Output port of Node 2
		Output port: Input table plus column with proportion missing (9 rows by 8 columns)

Node 4
Row Filter
	Filter criteria
	  Column to test: Proportion missing
	  Select Exclude rows by attribute value
	  Select; use range checking
	  Lower bound: 0.3
		Input port: Data table from the Output port of Node 3
		Output port: Rows with less than a 0.3 proportion of missing values retained (5 rows by 8 columns)

Node 5
Column Filter
	Filter:
	  Include: Var1–Var5
	  Exclude:
	  Count
	  Missing value count
		Input: Data table from the Output port of Node 4
		Output: Var1–Var5 for rows with less than 0.3 proportion of missing values (5 rows by 6 columns)

Node 6
Missing Value Column Filter
	Configuration:
	  Include: Var1–Var5
	  Missing value threshold (in %): 30.0%
		Input port: Data table from the Output port of Node 1
		Output port: Var2 and Var5 (other variables had more than 0.3 proportion of missing values) (9 rows by 2 columns)

Table 4.6 Output from Node 5 in the Fig. 4.3 workflow

RowID	Var1	Var2	Var3	Var4	Var5	Proportion of row missing
Row0	10	a	a	1	3.6	0.0
Row1	6	b	b	?	8.3	0.2
Row4	11	e	d	?	5.3	0.2
Row6	9	g	d	?	6.5	0.2
Row7	12	a	c	?	7.3	0.2

Table 4.7 Output from Node
6 in the Fig. 4.3 workflow

RowID	Var2	Var5
Row0	a	3.6
Row1	b	8.3
Row2	c	2.1
Row3	?	?
Row4	e	5.3
Row5	?	?
Row6	g	6.5
Row7	a	7.3
Row8	s	3.4

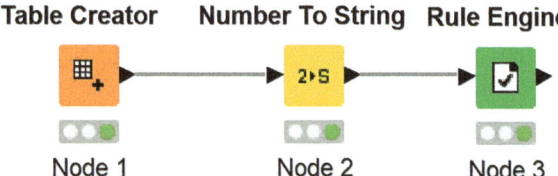

Fig. 4.4 Converting missing values to an indicator variable

Table 4.8 Node descriptions for the Fig. 4.4 workflow

Nodes	Descriptions	Input and output ports
Node 1 Table Creator	Create five variables Var1–Var5 with missing values.	Output port: Data table with created values
Node 2 Number to String	Options: Include: Var4 Exclude Var1 and Var5	Input port: Data table from the Output port of Node 1 Output port: Table with Var4 converted to string
Node 3 Rule Engine	Rule Editor Expression: MISSING $Var4$ => "Missing" TRUE => "Not missing" Check: Append Column: Var4 Missing	Input port: Data table from the Output port of Node 2 Output port: Input table plus added column with string variable indicating Var3 Missing or Not missing

The following options are available with the KNIME Missing Value node for
numeric variables:

- Do nothing.
- Remove the entire row that contains a missing value.
- Replace missing values with:

  – A fixed value specified by the user.
  – A summary statistic specified by the user: the maximum, minimum, mean, median, or rounded mean of the non-missing values.

Table 4.9 Output of Node 3 in Fig. 4.4 workflow

RowID	Var1	Var2	Var3	Var4	Var5	Var4 Missing
Row0	10	a	a	1	3.6	Not missing
Row1	6	b	b	?	8.3	Missing
Row2	?	c	?	?	2.1	Missing
Row3	4	?	?	3	?	Not missing
Row4	11	e	d	?	5.3	Missing
Row5	?	?	f	?	?	Missing
Row6	9	g	d	?	6.5	Missing
Row7	12	a	c	?	7.3	Missing
Row8	?	s	?	6	3.4	Not missing

- A linear interpolation between the previous and next encountered non-missing value in the column (useful for time series variables).
- The average value of the previous and next encountered non-missing value in the column (useful for time series variables).
- The mean of all non-missing values that are within a lookahead and look behind window specified by the user.
- The next or previous non-missing values (useful for time series variables).

The following options are available with the KNIME Missing Value node for string variables:

- Do nothing.
- Remove the entire row that contains a missing value.
- Replace missing values with:

  - A fixed value specified by the user.
  - The most frequent value (mode).
  - The next value.
  - The previous value.

Also, for categorical string variables, it is sometimes useful to explicitly consider the missing values as another level of the categorical variable.

Other strategies for imputing missing numeric values include:

- Regressing non-missing values of a variable on other variables in the data set and then using the regression equation to predict the values in the observations with the missing values.
- Using k nearest neighbors to find observations with similar values on the non-missing variables. Then use those observations to impute the missing values, for example, by taking the average of the non-missing values from k nearest neighbors. (The k Nearest Neighbors method is discussed in a later chapter.)

Additional methods for missing value imputation are available in R packages which can be incorporated into KNIME workflows, such as:

- MICE (Multivariate Imputation by Chained Equations) (Buuren et al., 2022).
- Amelia (A Program for Missing Data) (Honaker et al., 2021).
- VIM (Visualization and Imputation of Missing Values) (Templ et al., 2021).
- mlr (Machine learning in R) (Bischl et al., 2016).

4.6 Outliers

Outliers are defined as values of a numeric variable that are unusually different (i.e., extremely high or extremely low) from other variable values in a data set. While this may seem like an obvious criterion, whether a value is an outlier is not always easy to discern. Identifying and dealing with extreme outliers is important since these can cause problems for many analytics models.

While outliers can lead to biased or incorrect results, in some cases, outlying values may contain interesting or valuable information and should be carefully investigated. Identifying outliers is often the goal of analysis in areas like identifying credit card fraud, detecting voting irregularities, network intrusion, and climate studies.

Outliers are sometimes due to data entry errors, e.g., when an extra zero or two is added to or dropped from a value by mistake. In other cases, multiple outlying values can occur for various reasons. Domain knowledge helps explain whether an observation is an outlier or a valid value. For example, air travel, as measured by revenue passenger miles (RPM) by month, shows a sharp decline following 9/11/2000 due to the attacks on the World Trade Center and the Pentagon. This decline did not represent invalid data but rather legitimate values because of 9/11. In general, removing outliers without further investigation is not a good idea. If there is confidence that an outlier is an error, the observation can be removed if the fraction of cases with outliers is tiny.

A recommended approach for identifying outliers is using multiple techniques, including visual and quantitative tools (Aquinis et al., 2013). Two categories of outlier detection methods are available: univariate and multivariate techniques. Univariate techniques examine each variable in isolation, while multivariate techniques assess the distance of an observation from a central summary measure such as a centroid.

Univariate Outlier Detection Methods
Graphical methods, such as box and whisker plots, density plots, and histograms, can be used to identify potential outliers. The workflow in Fig. 4.5 provides an example of using multiple graphical techniques to identify potential outliers. Three characteristics of automobiles are used in the example: displacement, city MPG, and highway MPG.

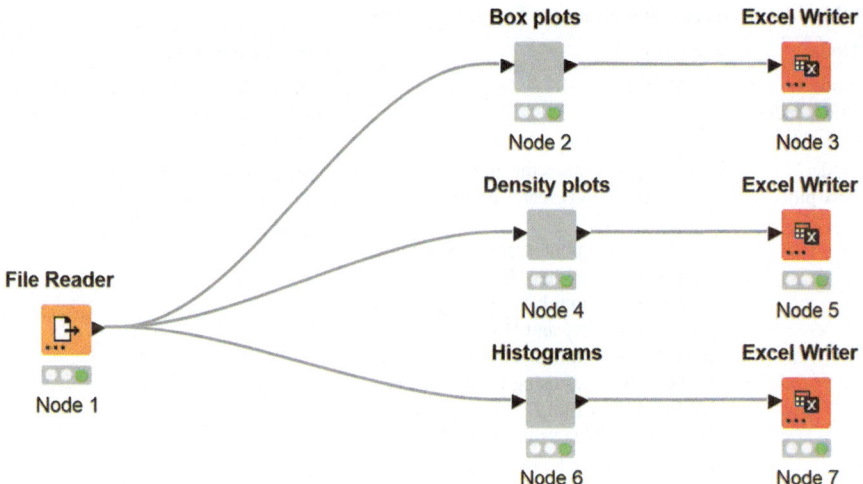

Fig. 4.5 Workflow for graphical detection of potential outliers

Each component (Nodes 2, 4, and 6) in Fig. 4.5 allows the user to select variables to include. The bandwidth[3] was set to 0.5 after some experimentation. In histograms in Node 6, the number of bins was set to 20. Node descriptions for Fig. 4.5 are provided in Table 4.10.

The graphs copied from the Excel files are shown in Figs. 4.6, 4.7, and 4.8.

The potential outliers are seen in the box plots and histograms for city miles per gallon and highway miles per gallon. Displacement is skewed to the right, but no obvious outliers are detected. The density plot for displacement indicates that there may be two overlapping distributions, one with relatively low displacement and the other with relatively high displacement. The shapes of the city and highway MPG density plots and histograms are also consistent with two overlapping distributions. It was hypothesized that the two underlying distributions for highway and city MPG were due to different displacement distributions. To further explore the data, the median value for displacement (3.3) was found using the Statistics node in KNIME and used to create two sets of histograms using the Row Splitter node, one for observations with displacement below the median and one for displacement above the median. (The workflow for this analysis is not shown.)

The resulting histograms for "City MPG" and "Highway MPG" are shown in Fig. 4.9. Two separate distributions are apparent, suggesting an interaction between miles per gallon and engine displacement. The highway MPG for smaller displacement engines still shows considerable right skew. The distributions for the higher displacement engines are more symmetric but still have several potential outliers.

Overlapping distributions may prove important as predictive models are developed with this data.

---

[3] Bandwidth in a density plot can vary from 0 (no smoothing) to ∞ (complete smoothing).

Table 4.10 Node descriptions for the Fig. 4.5 workflow

Nodes	Descriptions	Input and output ports
Node 1 File Reader	Read MPV.csv	<u>Output port:</u> Table with displ, cty, and hwy miles per gallon (234 rows by 3 columns)
Node 2 Box plots	<u>Component</u> that loops through selected input variables and creates boxplots of each variable Nodes in the component are shown below: 　　Node C1: Component Input 　　Node C2: Column Filter Configuration 　　Include displ, cty and Hwy 　　Node C3: Column List Start 　　Include variables using Flow Variables 　　Node C4: Box Plot 　　Include variables using Flow Variables 　　Node C5: Image to Table 　　Column name controlled by Flow Variable 　　Node C6: Renderer to Image 　　Node C7: Loop End (Column Append) 　　Node C8: Component Output 　　Exclude: column-filter	<u>Input port:</u> Data table from the Output port of Node 1 <u>Output port:</u> Table with box plots
Node 3 Excel Writer	Write Excel file to Box.xlsx Relative to current workflow	<u>Input port:</u> Data table from the Output port of Node 2
Node 4 Density plots	<u>Component</u> that loops through selected input variables and creates density plots of each variable Nodes in the component are shown below: 　　Node C1: Component Input 　　Node C2: Column Filter Configuration 　　Include displ, cty and Hwy 　　Node C3: Column List Loop Start. 　　Include variable using flow variables 　　Node C4: Double Configuration (outputs a double flow variable) 　　Control: The bandwidth is set interactively at 0.50 　　Node C5: Merge Variables 　　Merge the density variables with bandwidth 　　Node C6: 1D Kernel Density Plot 　　Node C7: Image to Table 　　Node C8: Loop End (Column Append) 　　Node C9: Component Output	<u>Input port:</u> Data table from the Output port of Node 1 <u>Output port:</u> Table with density plots
Node 5 Excel Writer	Write Excel file to Density.xlsx Relative to current workflow	<u>Input port:</u> Density plots from Node 4

(continued)

Table 4.10 (continued)

Nodes	Descriptions	Input and output ports

Node 6 Histograms	Component that loops through selected input variables and creates histograms of each variable
	Nodes in the component are shown below:
	  Node C1: Component Input
	  Node C2: Column Filter Configuration
	  Include displ, cty and Hwy
	  Node C3: Column List Loop Start.
	  Node C4: Integer Configuration
	  Node C5: Merge Variables
	  Merge the histogram variable with the number of bins
	  Node C6: Histogram
	  Parameters "num_bins", "title" and "cat" controlled by flow variables
	  Node C7: Image To Table
	  Column name controlled by flow variable
	  C8 Renderer to Image
	  Column name controlled by flow variable
	  Node C9; Loop end (Column Image Append)
	  Node C10: Component Output
		Input port: Data table from the Output port of Node 1
		Output port: Table with histograms plots

Node 7 Excel Writer	Write Excel file to Histogram.xlsx Relative to current workflow	Input port: Histograms from Node 7

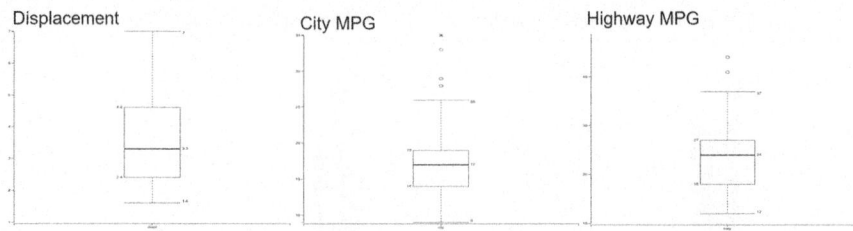

Fig. 4.6 Box plots for displacement, city and highway MPG

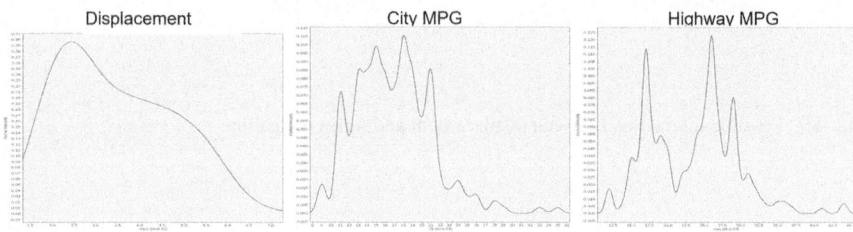

Fig. 4.7 Density plots for displacement, city and highway MPG

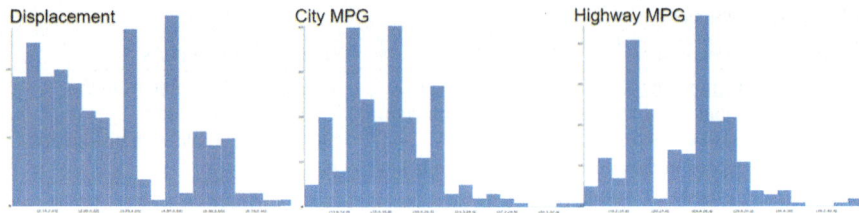

Fig. 4.8 Histograms for displacement, city, and highway MPG

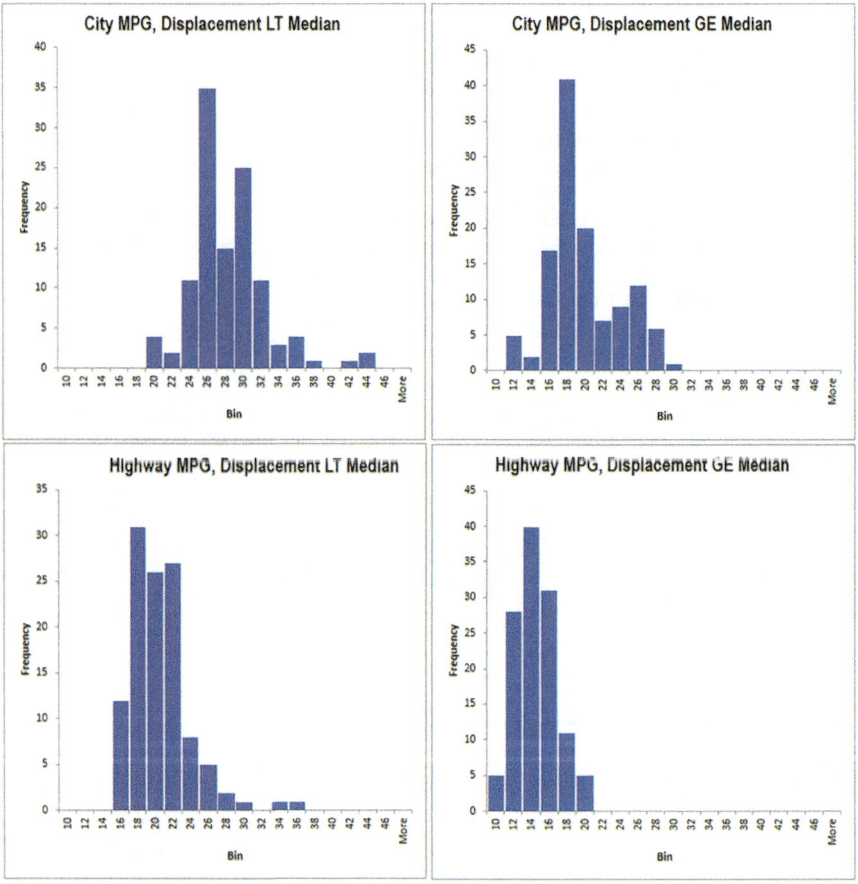

Fig. 4.9 Possible interaction between displacement and miles per gallon

Graphic methods are useful for visually inspecting variables, but other approaches involve statistical analyses. One common practice is for variables assumed to be normally distributed; a cutoff rule based on two or three standard deviations from the mean can be used to flag potential outliers. A value falling more than two standard deviations from the mean happens in about 5% of cases for a normal distribution. Values falling more than three standard deviations from the mean happen only in about 0.3% of cases. One weakness of this approach is that the standard deviations can be distorted by outliers, which can limit the identification of outliers. Also, a variable should be checked for symmetry before applying the standard deviation rule for detecting extreme values. Skewness can be used as an indicator of symmetry with the following rules of thumb can be used to assess symmetry:

- If the absolute value of skewness is less than and 0.5, a variable can be assumed to be at least approximately symmetric.
- If the absolute skewness value is between 0.5 and 1.0, a variable is moderately skewed and not quite symmetric.
- For skewness values greater than 1.0, a variable is skewed, and not symmetric.

Using a method based on non-parametric statistics is frequently thought to be a better way to identify potential outliers. The idea is based on Tukey's method, which uses upper and lower "fences" as limits to detect extreme values (Tukey 1977). The formulas are based on the interquartile range (IQR), which is the difference between the first (Q1) and third (Q3) quartiles of a variable or IQR = Q3 − Q1. If all variable values are ranked from highest to lowest, Q1 is the point below which between 25% of the values lie, and Q3 is the point below which 75% lie. A more general method depends on the cutoff value of k, set by the analyst. Values that fall more than k times the interquartile range above Q3 or k times the IQR below Q1 are considered extreme values.

$$\text{Lower cutoff} = Q1 - k(Q3 - Q1) \tag{4.1}$$

$$\text{Upper cutoff} = Q3 + k(Q3 - Q1) \tag{4.2}$$

A typical value for k in the upper and lower cutoffs is 1.5. The user can set the multiple of IRQ by right-clicking on the "Check extremes within k IQR" node and setting k. A number higher than 1.5 would identify fewer extreme values, while a lower number would identify more extreme ones.

The workflow in Fig. 4.10 with node descriptions in Table 4.11 uses the Numeric Outliers node, which implements the k IQR approach to identifying and treating outliers. This node allows the user to set the k value and to determine whether to delete rows with outliers or replace outlying values with the closest permitted value.

The results from running the Fig. 4.10 workflow are shown in Table 4.12. Using the ±1.5 IQR no outliers were found for displacement, 5 for city MPG, and 3 for highway MPG. The lower and upper bounds used to test for outliers for each variable are shown in columns four and five of Table 4.12.

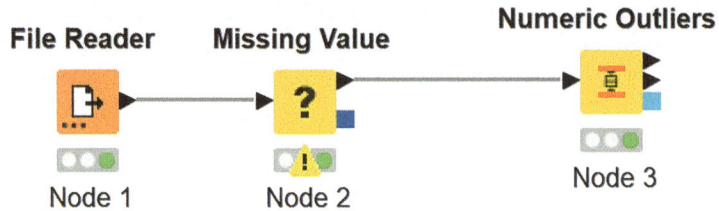

Fig. 4.10 Workflow to flag outliers

Table 4.11 Node descriptions for the Fig. 4.10 workflow

Nodes	Descriptions	Input and output ports
Node 1 File Reader	Read MPG.csv	Output port: Table with 236 rows and 3 columns
Node 2 Missing Value	Default: Number (double): Remove Row Number (integer): Remove Row	Input port: Data table from the Output port of Node 1 Output port: Table with 234 rows and 3 columns after rows with missing values deleted
Node 3 Numeric Outliers	Settings for the node Outlier Selection: Include displ, city, and hwy General Settings: Interquartile range multiplier (k)· 1 5 Quartile Calculation: Full data estimate using R_4 Outlier Treatment: Apply to: All outliers Treatment option: Replace outlier values Replacement Strategy: Closest permitted value	Input port: Data table from the Output port of Node 2 Upper output port: Treated table Middle output port: Summary table with number of observations for each variable, number of outliers, the lower and upper bounds created using the IQR rule

Table 4.12 Outliers identified in the MPG file

Outlier column	Member count	Outlier count	Lower bound	Upper bound
displ	234	0	−0.9	7.9
cty	234	5	6.5	26.5
hwy	234	3	4.5	40.5

Detecting Multivariate Outliers
To this point, the discussion has focused on univariate outliers. Even if no univariate outliers are found, outliers may be present that can only be detected using multivariate methods. For example, the scatterplot (Fig. 4.11) shows an outlier (circled).

Figure 4.11 clearly shows an outlying observation, yet the box plots for each variable, shown in Fig. 4.12, do not indicate the presence of an outlier because the outlying value is within the range of both the x and y variables.

The Mahalanobis Distance (MDist) can be used to identify multivariate outliers. A detailed discussion of the formula and a discussion of the usage is in Cansiz (2020). KNIME does not have a node for computing MDist, so an R Snippet was used, as shown in Fig. 4.13. The node descriptions are in Table 4.13. Table 4.14 shows that observation number 6 correctly identified as a multivariate outline with an MDist of 9.19.

Handling Outliers
After identifying potential outliers, the next step is deciding what to do about them. Outlying values could be legitimate, correct values, so routinely deleting what appear to be outliers is not a recommended practice. In fact, in some cases, detecting outliers is the object of analysis in applications such as intrusion detection in computer systems, determining fraud in financial activity, or detecting unusual patterns in medical data (Aggarwal, 2017). It is up to the analyst to decide whether outliers are legitimate or if they represent errors.

If the outliers are not determined to be the results of data entry, measurement, or other data errors, they should not be routinely removed. Removing outliers is

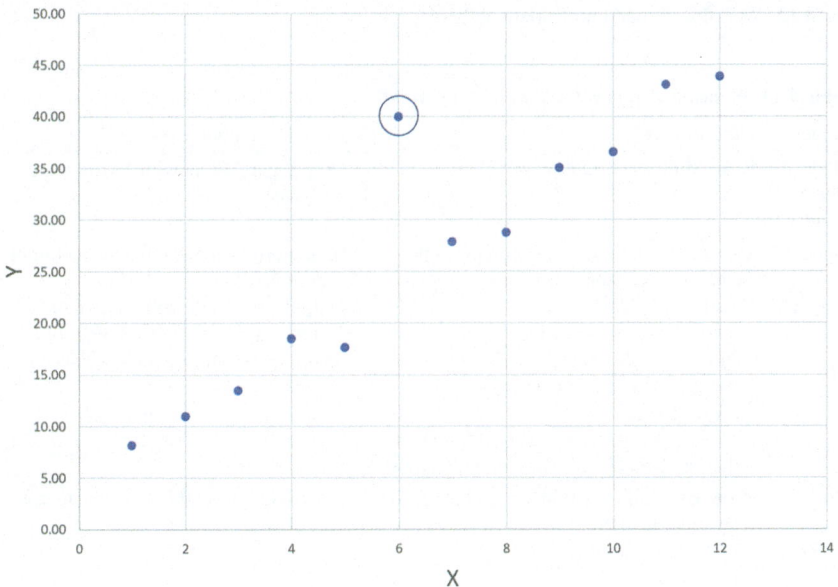

Fig. 4.11 Example of a multivariate outlier

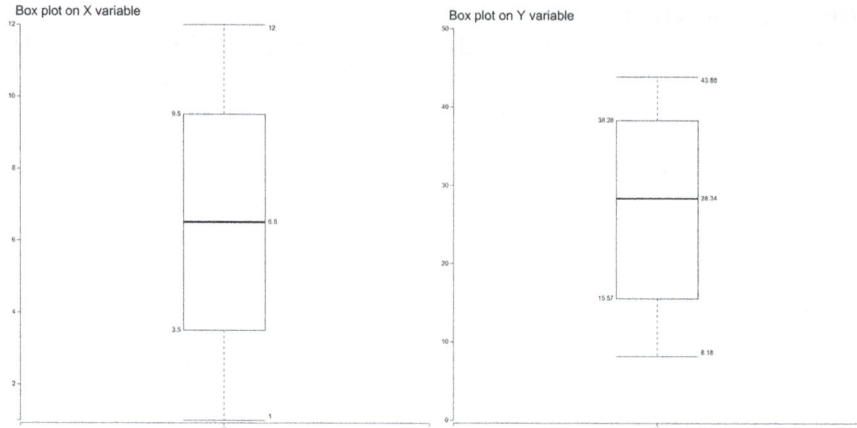

Fig. 4.12 Box plots on X and Y variables

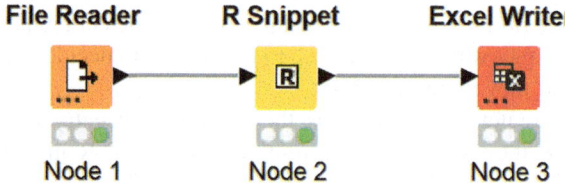

Fig. 4.13 Workflow to find multivariate outliers

Table 4.13 Node descriptions for the Fig. 4.13 workflow

Nodes	Descriptions	Input and output ports
Node 1 File Reader	Read MultivariateOutlier.csv	Output port: Table with 12 rows and 2 columns

Node 2 R Snippet	The R code in this snippet computes the Mahalanobis Distance of the x y points R code for Node 2:
	 `data <- knime.in`
	 `mah <- as.data.`
	 `frame(mahalanobis(data.`
	 `matrix(data),`
	 `colMeans(data),cov(data)))`
	 `knime.out <- cbind(data,mah)`
		Input port: Data table from the Output port of Node 1
		 Output port: Table with input table of 12 rows and 2 columns of input plus a column with Mahalanobis distances

Node 3 Excel Writer	Write Excel file to MultivariateDistance.xlsx relative to the current workflow data area	Input port: Data table from the Output port of Node 2

Table 4.14 Mahalanobis
distances for each
observation

X	Y	Mahalanobis distance
1	8.18	2.34
2	11.02	1.61
3	13.46	1.11
4	18.52	0.48
5	17.69	0.99
6	40.00	**9.19**
7	27.89	0.04
8	28.79	0.61
9	35.05	0.48
10	36.56	1.12
11	43.15	1.62
12	43.88	2.41

unlikely to improve predictive performance, and this can be checked by running predictive analyses with and without the outliers (or with the outliers adjusted to permitted values).

If it is deemed necessary to treat or remove outliers, the KNIME node Numeric Outliers detects and provides several options to treat any identified outliers. The treatment options are:

- Remove each row containing an outlier (although this should only be done with careful consideration).[4]
- Replace each outlier with a missing value indicator.
- Replace each outlier with the closest value within the permitted interval (known as Winsorizing) (Tukey, 1962).

Sometimes a log or power transformation will reduce variability in a predictor and lessen or remove outliers. The Math node in KNIME can be used to create transformations.

An approach to dealing with skewness is binning,[5] which involves dividing a variable into groups of values. The KNIME node Numeric Binner can define and develop bins; the numeric input variable is transformed into a string of ordinal values. Of course, detail is lost in binning, but this is sometimes a useful approach if there are extreme outliers. The analyst needs to specify how the bins are formed by selecting the number of bins and the range of values for each bin. Two ways to set the boundaries or edges of the bins once the number of bins is set are:

---

[4] Outliers should not be routinely deleted because they are not necessarily bad observations. In some cases, they can be quite informative. On the other hand, if the outlier appears to be a recording or measurement error or is otherwise invalid, then the outlier can be discarded. It is important to have domain knowledge to make such judgments.

[5] Binning is also referred to as "discretization" or "bucketing."

- Make each bin contain an equal (or approximately equal) number of cases. The advantage of this method is that the balanced bins can handle highly skewed data.
- Create bins that have equal width values. The advantage of this approach is that it preserves the distribution of the variable.

Both ways of setting up the bins result in creating an ordinal variable.

4.7   Feature Engineering

Feature engineering refers to steps to improve the accuracy of predictive models through data transformations, constructions of new variables from existing variables, or acquiring additional variables to enhance model performance. Despite its technical connotation, feature engineering is more of an art than a science learned through experience.[6]

The first step in feature engineering is exploring the data using descriptive statistics and graphical representations. Exploratory data analysis is an approach/philosophy for data analysis that employs various visual and descriptive statistical techniques to provide insight into a data set.

Exploratory analysis is open-ended since the goal is to generate clues about what is happening, which can then open new avenues of investigation ("hypothesis generation" rather than the more formal "hypothesis testing" featured prominently in many books on statistics). Exploration of the data can also support selecting appropriate modeling techniques, finding underlying structures in the data, and assessing the cleanliness of the data

Since exploration does not involve a fixed set of steps or tools, the software should be easy to use and flexible. Both visual and statistical tools are typically used. Visual techniques for data exploration include the following commonly used graphs:

- Scatter plots
- Bar plots
- Histograms
- Box and whisker plots
- Density plots
- Bubble charts

Commonly used descriptive statistics include:

- Mean, median, mode
- Variance, standard deviation

---

[6] Terms such as independent variables, predictors, factors, and inputs are typically used by statisticians, while the term *features* is more prevalent in the context of machine learning. In addition, some authors distinguish between "raw" variables and "features," with the latter referring to constructed variables.

- Range, Max, Min
- Correlations and covariances
- Frequencies and percentages
- Percentiles and quartiles

Transformations to Reduce Skewness
Some data mining and predictive techniques work better with at least approximately symmetric variables. For example, models can be overly sensitive to the heavy tails in distributions. While predictor values in machine learning models do not have to be perfectly symmetrical, highly skewed predictors can distort or bias model performance. One quick way to identify non-symmetric distributions is to examine each variable for extreme skewness.

Detecting skewness is usually straightforward, using histograms and density plots. Skewness can be reduced by using transformations. For positive or right skew (where the tail of the distribution is heavier on the right side), the following transformations can be applied (listed in order of the strength) to a variable x: where k is a large enough positive number to ensure that the log and square root are applied only to numbers greater than 0.0.

- $(k + x)^{1/2}$
- $(k + x)^{1/3}$
- $\log(k + x)$
- $1/(k + x)$

The left panel of Fig. 4.14 shows a histogram with a strong positive skew. The center panel shows the histogram for a square root transformation, and the right panel shows the result for a log transformation. The square root transformation created a more symmetric distribution, while the log transformation created a slightly negative skew. Which transformation to use is an empirical question and depends on the data.

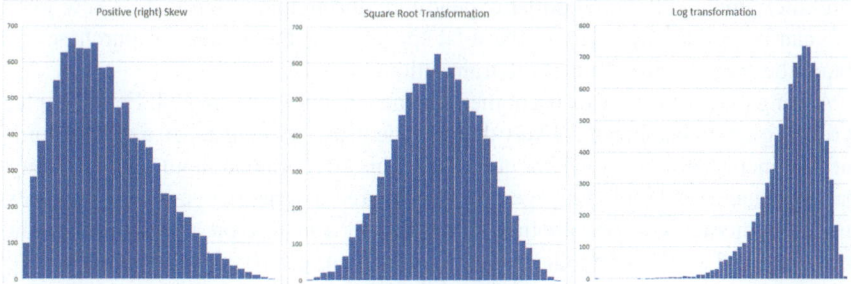

Fig. 4.14 Transformations for positive skew

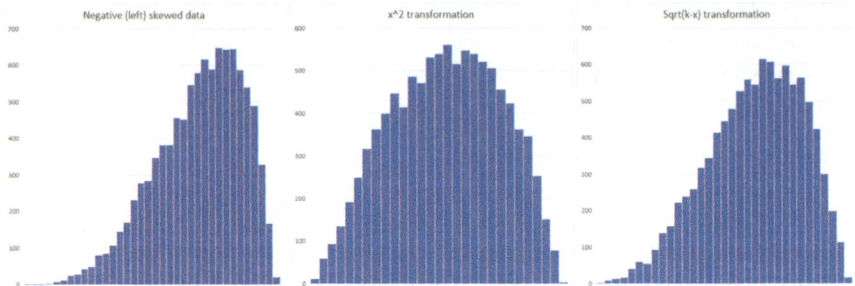

Fig. 4.15 Transformations for negative skew

For negative or left skew (where the tail of the distribution is heavier on the left side), the following transformations can be applied (listed in order of strength) to a variable x:

- $(k - x)^{1/2}$, where k is large enough to make $(k - x)$ positive for all observations.
- x^2
- x^3
- x^n where n is greater than 3.

The left panel of Fig. 4.15 shows a histogram with a strong negative skew. The center panel shows the histogram for a transformation to the power of 2. The right panel shows the result for the square root of $(k - x)$, with k set large enough to make all entries positive. The square root transformation again worked well. Which approach works best with an observed variable is an empirical question usually explored by trying various transformations.

Deriving New Variables

New variables can be created using two or more variables already in the data set. For example, a data set on customer purchases may show the number of purchases made in a timeframe and the total dollar expenditure during that timeframe. A new variable can be derived by dividing the total spending by the number of purchases to obtain the average size of purchase transactions.

Adding polynomial terms is another way new features are created. For example, in relating age to income, it is likely that income increases with age to a certain point but then increases more slowly or decreases. This effect can be captured by creating the age-squared and including age and age-squared in a predictive model. Features may also interact with one another. The effect of a feature on a prediction might depend upon the value of another feature. Interactions can be created by multiplying two continuous variables.

Date and time variables read as strings also need to be converted into multiple fields, e.g., a date in the form "1/1/2000" to Month = 1, Day of the month = 1, and Year = 2000.

4.8   Example of Using KNIME for Data Preparation

To illustrate some of the data preparation tasks using KNIME, a slightly modified data set on red wine from UCI will be used[7]. The workflow in Fig. 4.16 uses the Statistics node in KNIME to obtain descriptive stats on the data set. Node descriptions for the workflow are in Table 4.15.

Table 4.16 shows the results of the analysis. The following variables have high skewness values: residual sugar (4.54); chlorides (5.68); free sulfur dioxide (1.25); total sulfur dioxide (1.52); and sulfates (2.43). These variables are moderately skewed: fixed acidity (0.98); volatile acidity (0.67); and alcohol (0.86).

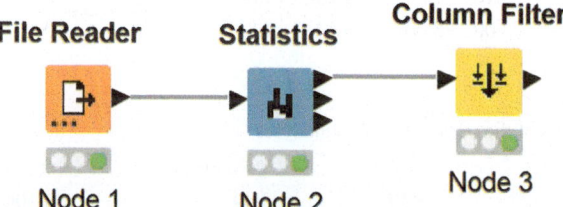

Fig. 4.16 Workflow for descriptive statistics

Table 4.15 Node descriptions for the Fig. 4.16 workflow

Nodes	Descriptions	Input and Output ports
Node 1 File Reader	Read RedWineData	Output port: KNIME table with 1,999 rows and 12 columns
Node 2 Statistics	Place all variables in the Include area Check box: Calculate median values	Input port: Data table from the Output port of Node 1 Upper output port: Statistics table with 12 rows and 16 columns for all numeric variables Middle Output port: Not used Lower output port: Not used
Node 3 Column Filter	Place the following columns in the Include area: Min Max Mean Std. deviation Skewness No. missing Median Place the remaining columns in the Exclude area	Input port: Data table from the Upper output port of Node 2 Output port: Statistics table with 12 rows and 7 columns for all numeric variables

---

[7] https://archive.ics.uci.edu/dataset/186/wine+quality

Table 4.16 Descriptive statistics on the red wine data set

Variable	Min	Max	Mean	Std. dev.	Skewness	# missing	Median
fixed acidity	4.60	15.90	8.32	1.74	0.98	0	7.90
volatile acidity	0.12	1.58	0.53	0.18	0.67	2	0.52
citric acid	0.00	1.00	0.27	0.19	0.32	0	0.26
residual sugar	0.90	15.50	2.54	1.41	4.54	0	2.20
chlorides	0.01	0.61	0.09	0.05	5.68	1	0.08
free sulfur dioxide	1.00	72.00	15.87	10.46	1.25	0	14.00
total sulfur dioxide	6.00	289.00	46.49	32.90	1.52	1	38.00
density	0.99	1.00	1.00	0.00	0.07	2	1.00
pH	2.74	4.01	3.31	0.15	0.19	0	3.31
sulfates	0.33	2.00	0.66	0.17	2.43	1	0.62
alcohol	8.40	14.90	10.42	1.07	0.86	0	10.20
quality	3.00	8.00	5.64	0.81	0.22	0	6.00

Table 4.17 Results of the outlier detection analyses

Column	# obs	# outliers	% outliers
fixed acidity	1592	49	3.10%
volatile acidity	1592	19	1.20%
citric acid	1592	1	0.10%
residual sugar	1592	154	9.70%
chlorides	1592	112	7.00%
free sulfur dioxide	1592	30	1.90%
total sulfur dioxide	1592	55	3.50%
density	1592	45	2.80%
pH	1592	35	2.20%
sulfates	1592	59	3.70%
alcohol	1592	13	0.80%
quality	1592	28	1.80%

Five variables have one or two missing values. Since there were only a few missing values, the Missing Value node in KNIME will be used in the following workflows to replace the missing data with the median values for the respective variables.

Next, the workflow in Fig. 4.10 was used with the red wine data to identify potential outliers. The results of checking for outliers using the 1.50 IQR rule are in Table 4.17 and show a substantial number of outliers for several of the variables. Note that the variables with the largest percentage of outliers ("residual sugar" and "chlorides") had the highest skew values reported in Table 4.16 (skews of 4.54 and 5.68).

Next, box plots, density plots, and histograms were created for the residual sugar since it had the largest number of 1.5 IQR outliers. The workflow in Fig. 4.5 was modified to read the Red Wine data. The charts, shown in Fig. 4.17, clearly show the outliers and skewness of these variables.

Dealing with Outliers and Extreme Variables in This Example

The variable "residual sugar" will be used to illustrate several alternative ways of dealing with an outlier. Table 4.17 and Fig. 4.17 showed that "residual sugar" had many outliers. Using the Math node in KNIME, several transformations of "residual sugar" were tried using the workflow in Fig. 4.18. The transformations tried included (residual sugar)$^{1/2}$, (residual sugar)$^{1/3}$, log(residual sugar), and 1/(residual sugar). Only the inverse transformation 1/(residual sugar) created a more symmetric distribution. Node descriptions are in Table 4.18.

The number of outliers was also obtained in the Fig. 4.18 workflow, with results shown in Table 4.19. While the inverse of the "residual sugar" variable still evidences many outliers when measured with the 1.5 IQR rule, the histogram shown in Fig. 4.19 indicates a reasonably symmetric distribution for the transformed variable. Whether or not this transformation results in improved predictive performance is an empirical question. Ideally, predictive analyses using the non-transformed and transformed variable will be run the results compared.

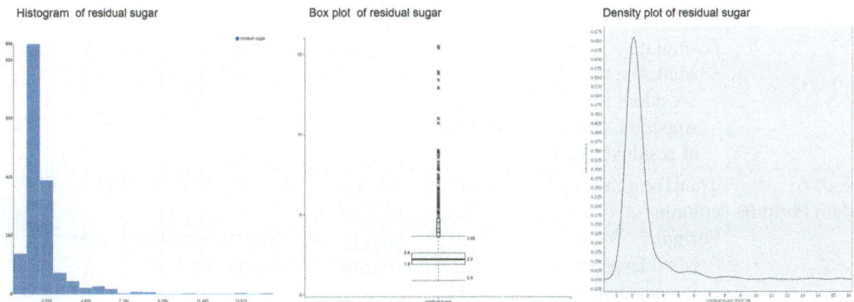

Fig. 4.17 Histogram, box plot, and density plot, for residual sugar variable

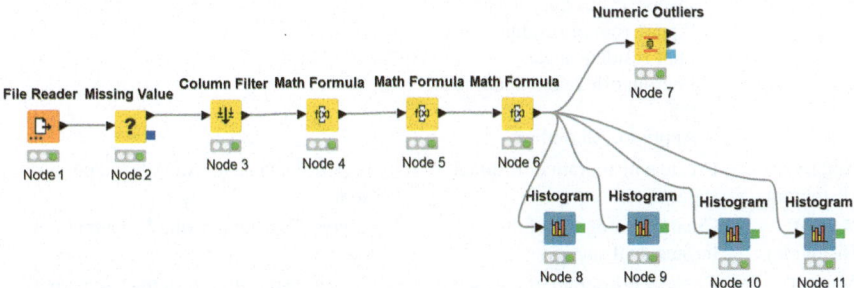

Fig. 4.18 Workflow to assess transformations of residual sugar

Table 4.18 Node descriptions for the Fig. 4.18 workflow

Nodes	Descriptions	Input and output ports
Node 1 File Reader	Read RedWineData	Output port: KNIME table with 1999 rows and 12 columns
Node 2 Missing Value	Remove rows with missing values	Input port: Data table from the Output port of Node 1 Output port: Data table with column with any missing rows removed
Node 3 Column Filter	Remove all variables except for residual sugar	Input port: Data table from the Output port of Node 2 Output port: Data table with residual sugar

Node 4
Math Formula
	Transform the "residual sugar" column
	Formula:
	    Math Expression:
	    1/($residual sugar$)
	    Append Column: Inverse of residual sugar
		Input port: Data table from the Output port of Node 3
		Output port: Data table with added column "Inverse of residual sugar"

Node 5
Math Formula
	Transform the "residual sugar" column
	Formula:
	    Math Expression:
	    $residual sugar$^.5
	    Append Column: Square root of residual sugar
		Input port: Data table from the Output port of Node 4
		Output port: Data table with added column "Square root of residual sugar"

Node 6
Math Formula
	Transform the "residual sugar" column
	Formula:
	    Math Expression:
	    ln($residual sugar$)
	    Append Column: ln of residual sugar
		Input port: Data table from the Output port of Node 5
		Output port: Data table with an added column "ln of residual sugar"

Node 7
Numeric
Outliers
	Outlier Settings:
	    Include:
	    residual sugar
	    Inverse of residual sugar
	    Square root of residual sugar
	    ln of residual sugar
	    Interquartile range multiplier:
	    0.5
	    Apply to: All outliers
		Input port: Data table from the Output port of Node 6
		Output port: Not used here

Node 8 Histogram	Create a histogram of residual sugar	Input port: Data table from the Output port of Node 6
Node 9 Histogram	Create a histogram of the ln($residual sugar$)	Input port: Data table from the Output port of Node 6
Node 10 Histogram	Create a histogram of residual sugar$^.5	Input port: Data table from the Output port of Node 6 Output port: Data table with added column "Square root of residual sugar"
Node 11 Histogram	Create a histogram of the Inverse of residual sugar	Input port: Data table from the Output port of Node 6

Table 4.19 Outliers in residual sugar variable without transformation plus square root, natural log, and inverse transformations

Column	# obs	# outliers	% outliers
residual sugar	1599	155	9.70%
Square root(residual sugar)	1599	144	9.00%
LN(residual sugar)	1599	121	7.60%
Inverse(residual sugar)	1599	70	4.40%

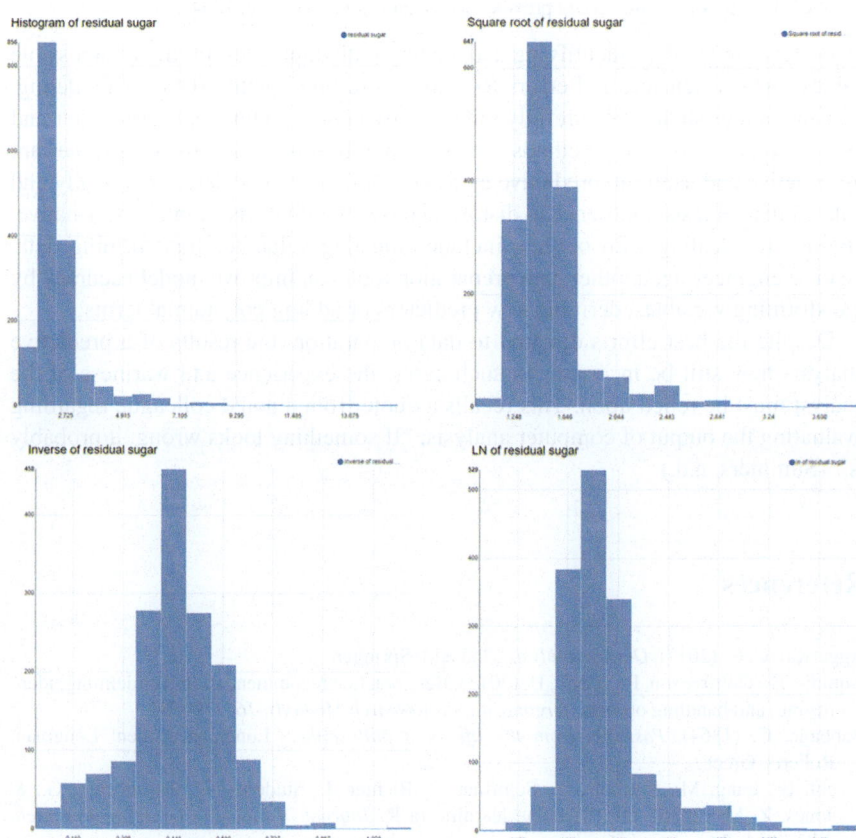

Fig. 4.19 Distribution of residual sugar untransformed and transformed three ways

4.9   Summary

Data preparation can be tedious, but it is critical that careful attention and time is devoted to cleaning and preparing the data. The performance and usefulness of predictive models depend on having the right data in the right form. The GIGO concept ("garbage in, garbage out") was noted by Charles Babbage, who is credited with the first design for a programmable computer:

> On two occasions, I have been asked, "Pray, Mr. Babbage, if you put into the machine wrong figures, will the right answers come out?" I am not able rightly to apprehend the kind of confusion of ideas that could provoke such a question. (Babbage, 1864)

Data cleaning involves identifying and dealing with duplicates, outliers, impossible values, inconsistencies, and errors to help ensure high-quality data for modeling. Missing values should be carefully managed by understanding why they occur and using imputation or other methods if appropriate. In some cases, missing values are informative and useful in predictive analyses. Outliers can be detected visually and with statistical tests. Outliers can distort analyses, but they also can be informative. Options for dealing with outliers include adjusting values or transforming data. Feature engineering, another data preparation tool, can improve model accuracy by transforming variables, deriving new predictors or adding polynomial terms.

Despite the best efforts directed to data preparation, the results of a predictive analysis may still be incorrect. In such cases, the experience and wariness of the analyst must be relied upon. This recalls a quote from a noted colleague regarding evaluating the output of computer analysis: "If something looks wrong, it probably is." (Summers, n.d.)

References

Aggarwal, C. C. (2017). *Outlier analysis* (2nd ed.). Springer.

Aquinis, H., Gottfredson, R., & Joo, H. (2013). Best-practice recommendations for defining, identifying, and handling outliers. *Organizational Research Methods, 16*, 270–301.

Babbage, C. (1864). *Passages from the life of a philosopher*. Longman, Green, Longman, Roberts, Green.

Bischl, B., Lang, M., Kotthoff, L., Schiffner, J., Richter, J., Studerus, E., Casalicchio, G., & Jones, Z. M. (2016). mlr: Machine learning in R. *Journal of Machine Learning Research, 17*(170), 1–5.

Buuren, S., et al. (2022). *Package "mice"*. https://cran.r-project.org/web/packages/mice/mice.pdf. Accessed 29 July 2023.

Cansiz, S. (2020). *Mahalanobis distance and multivariate outlier detection in R*. https://builtin.com/data-science/mahalanobis-distance. Accessed 29 July 2023.

Honaker, J., King, G., & Blackwell, M. (2021). *Package "Amelia"*. https://cran.r-project.org/web/packages/Amelia/Amelia.pdf. Accessed 29 July 2023.

Little, R. J., & Rubin, D. B. (1989). The analysis of social science data with missing values. *Sociological Methods & Research, 18*(2–3), 292–326.

Rubin, D. B. (1976). Inference and missing data. *Biometrika, 63*(3), 581–590.

Summers, J. (n.d.). *If something looks wrong, it probably is wrong.*

Templ, M., et al. (2021). *Visualization and imputation of missing values.* https://cran.r-project.org/web/packages/VIM/VIM.pdf. Accessed 29 July 2023.

Tukey, J. (1962). The future of data analysis. *Annals of Mathematical Statistics, 33*(1), 1–67.

Tukey, J. (1977). *Exploratory data analysis.* Addison-Wesley.

Chapter 5
Dimensionality Reduction

The number of predictor variables in a data set could be hundreds, thousands, or more in applied situations.[1] This might not seem to be a problem – it is sometimes thought that "more data is better." Even with a moderately large number of variables, creating thousands of potential predictors is possible. For example, one study predicting the onset of personal bankruptcy among users of credit cards began with 255 potential predictors. Sums, differences, products, and ratios of the 255 initial variables resulted in over 67,000 potential predictors.

5.1 Problems with Large Numbers of Variables

In practice, having too many variables, especially irrelevant or redundant variables, can reduce the effectiveness of a business analytics project. The "curse of dimensionality" is a label applied to the exponential increase in the number of observations needed to maintain the accuracy of a predictive model as the number of predictors increases. The Hughes phenomenon (Hughes, 1968) states that with a fixed number of data points, the predictive performance of a model increases at first as the number of predictors increases but then decreases as further predictors are added.

Several simulations and experiments with analytic techniques have shown that some are sensitive to having irrelevant variables tossed into the mix. So, if random variables are included in the mix of potential predictors, this can cause the predictive ability of a model to degrade. The exact effect depends upon the technique used, but many predictive techniques suffer from the same problem. More surprisingly, it has also been found that even having too many relevant variables can also reduce overall accuracy.

---

[1] In articles, books, and business analytics and machine learning presentations, predictors might be called variables, features, or attributes.

© The Author(s), under exclusive license to Springer Nature Switzerland AG 2023   85
F. Acito, *Predictive Analytics with KNIME*,
https://doi.org/10.1007/978-3-031-45630-5_5

As dimensionality increases, the data set becomes sparser. It has been discussed in an earlier chapter that validating a predictive model usually requires that the available data be split into two or more subsets, one used for training a model and the other for testing and validation. The implicit assumption is that each of these subsets will have the same pattern of values for variables and all combinations of the variables. Every new dimension increases the chance that the training and other subsets will not be comparable. This means more samples are needed in the training, test, and validation data sets to be confident of the model's generalizability.

Having too many variables can also cause other undesirable effects. For one thing, computer processing time increases as the number of variables increases. While computer processing time is not likely a problem when developing a model, it can be a concern when the model is deployed where real-time, instantaneous predictions are needed.

Another problem with large numbers of variables is that the model becomes difficult to maintain. When a predictive model is deployed into production, values for all the predictors must be obtained, stored, and cleaned. Furthermore, a predictive model becomes more complex with many predictors, and it is more likely that unusual or strange relationships are found that will be difficult to understand or explain.

Finally, with many predictors, it becomes highly likely that redundant or nearly redundant variables are included. This can cause instability in the model as model coefficients appear to bounce around as different data subsets are analyzed.

Often there are variables in a data set that have no relationship to the target. For example, customer ID numbers can typically be dropped.[2] In certain predictive models, variables such as race or ZIP code must be removed due to regulatory concerns.

5.2 Approaches to Dimension Reduction

Three general approaches to reducing the dimensionality of a data set are available:

- Manually removing variables with excessive missing values, high inter-variable correlations, zero or near-zero variance, or low correlation with the target variable.
- Using algorithms designed to comb through variables and pick the most predictive ones. These are generally referred to as wrapper methods.
- Reducing dimensionality through transforming and combining predictors. Forming combinations of variables can be done by the analyst using domain knowledge or by using an automatic approach with principal components analysis.

---

[2] ID numbers sometimes contain implicit information such as category or date. However, there are usually more direct ways to include such information.

All the strategies discussed here are heuristic and will not guarantee that the best subset is created (Silipo et al., 2014).

Arbitrarily removing any variables when developing a predictive model is not good practice. However, it is sometimes useful to manually remove one or more possible predictors in cases where there are many predictors, but this should be done carefully. If, after removing variables using the methods discussed below, the predictive model's performance is quite good and satisfies the objectives of the analysis project, then it is likely that little harm has been done. A better approach would be to run the analysis with all predictors and then with the reduced set to evaluate the degradation in predictive power. With these caveats in mind, descriptive statistics can help reduce the number of dimensions.

Highly correlated variables can be removed as a means of predictor selection. If two variables are highly correlated, one can be dropped. The choice of which one of the pair to drop should be based on the analyst's judgment about which is likely to be more meaningful. However, there are situations when including two redundant variables can result in more accurate predictions than either variable alone, so care and intuition are needed.

The distribution of each potential predictor can be examined to identify those that can be eliminated. Variables with many missing values may be candidates for pruning, but it must be recognized that missingness might help predict the target variable. Domain knowledge can guide this approach. Suppose the analyst suspects a missing value on a particular variable is informative about the target. In that case, a binary indicator variable can be created to flag such cases.

If the distribution of a potential predictor shows that the variable is constant or nearly constant (e.g., with very small variance), the variable might be dropped. Again, domain knowledge and judgment should guide this decision.

Removing Highly Correlated Variables

A workflow to select variables below a correlation threshold is in Fig. 5.1. Predictive accuracy is almost always reduced if even highly correlated variables are removed. The problem with correlated predictors is that it can be impossible to identify the effect or importance of specific variables. The model also can become very sensitive to changes in the number of observations or even unstable.

To help identify highly correlated variables, the workflow in Fig. 5.1 uses simulated data with 1000 observations on ten variables at various levels of correlation. Node descriptions are provided in Table 5.1.

Table 5.2 shows the correlation matrix among the eight simulated variables. For each column in the correlation matrix, the number of correlations above the threshold value is counted. The column with the most correlated columns above the threshold is chosen to "survive" and all correlated columns are filtered out. This procedure is repeated until no more columns can be identified. While the problem of finding a minimum set of columns to satisfy the constraints is difficult to solve analytically, this method applied here is known to be a good approximation. For example, if the threshold is set to 0.70, only variables V1, V3, V4, and V5 remain, and the rest are filtered out.

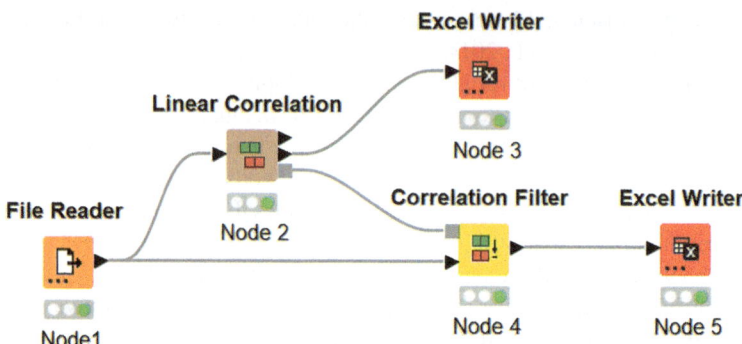

Fig. 5.1 Workflow to remove highly correlated variables

Table 5.1 Node descriptions for the Fig. 5.1 workflow

Nodes	Descriptions	Input and output ports
Node 1 File Reader	Read file: CorrelatedSimulatedVariables.csv	Output port: Data table with 1000 rows and 8 columns
Node 2 Linear Correlation	Options: Include all continuous variables	Input port: Data table from the Output port of Node 1 Upper output port: Not used Middle output port: 10 × 10 correlation matrix Lower output port: Correlation model
Node 3 Excel Writer	Settings: Write to: Relative to Current workflow area File: CorrelationMatrix.xlsx	Input port: Data table from the Middle output port of Node 2
Node 4 Correlation Filter	Settings: Slider to set threshold for removing highly correlated variables After setting slider, click on Calculate to show the number of columns included and excluded	Upper input port: Data table from Lower output port of Node 2 Lower input port: Data table from the Output port of Node 1 Output port: Filtered data (1000 rows by 4 columns)
Node 5 Excel Writer	Settings: Write to: Relative to Current workflow area File: FilteredDataSet.xlsx	Input port: Data table from the Output port of Node 4

Table 5.2 Correlation matrix from workflow in Table 5.1

Variable	V1	V2	V3	V4	V5	V6	V7	V8	V9	V10
V1	1.000	0.794	0.491	0.204	0.035	0.772	0.483	0.763	0.772	0.482
V2	0.794	1.000	0.405	0.314	0.031	0.972	0.400	0.960	0.971	0.399
V3	0.491	0.405	1.000	0.617	0.508	0.393	0.964	0.387	0.393	0.963
V4	0.204	0.314	0.617	1.000	0.269	0.301	0.597	0.296	0.302	0.596
V5	0.035	0.031	0.508	0.269	1.000	0.026	0.478	0.025	0.025	0.476
V6	0.772	0.972	0.393	0.301	0.026	1.000	0.386	0.999	0.999	0.385
V7	0.483	0.400	0.964	0.597	0.478	0.386	1.000	0.380	0.399	0.999
V8	0.763	0.960	0.387	0.296	0.025	0.999	0.380	1.000	0.998	0.379
V9	0.772	0.971	0.393	0.302	0.025	0.999	0.399	0.998	1.000	0.398
V10	0.482	0.399	0.963	0.596	0.476	0.385	0.999	0.379	0.398	1.000

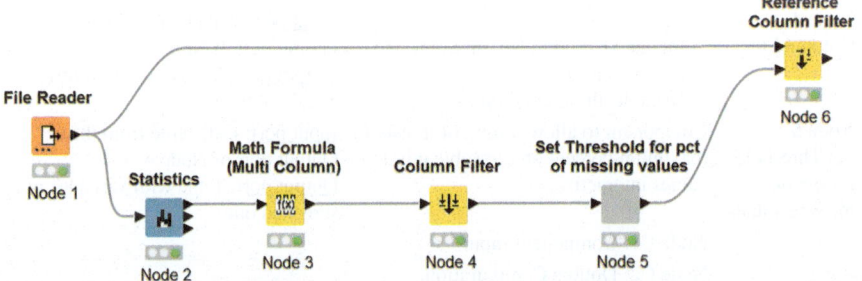

Fig. 5.2 Workflow to screen variables by percent of missing values

Removing Variables with an Excessive Number of Missing Values

Variables can also be filtered if the number or percentage of missing values exceeds a threshold set by the analyst. Figure 5.2 shows the workflow for this process; the node descriptions are in Table 5.3.

Before removing variables based on missingness, imputation techniques should be investigated to determine the effect of substituting the mean, median, or other values on a predictive model.

Removing Variables with Zero or Near-Zero Variance

If a predictor has zero variance, it is a constant and thus will not have any predictive power. Some algorithms, such as regression, will fail if such a variable is included. It is best to remove constants from the predictive set.

There is also the issue of potential predictors with very low or near-zero variance. The reasoning is that a feature with low variance cannot explain much of the variance in the response variable. The magnitude of variance is affected by the numerical range of a variable, so the usual practice is to first normalize all predictors to a range of 0–1.[3]

---

[3] Standardization to zero mean and unit standard deviation should *not* be done since this will make all variances equal.

Table 5.3 Node descriptions for the Fig. 5.2 workflow

Nodes	Descriptions	Input and output ports
Node 1 File Reader	Read file: MissingValueSimulatedData. csv	Output port: Data table with 1000 rows and 6 columns
Node 2 Statistics	Options: Include all continuous variables	Input port: Data table from the Output port of Node 1 Upper output port: Statistics table with six rows and 16 columns Middle output port: Not used Lower output port: Not used
Node 3 Math Formula (Multi Column)	Math Expression: Include: No. missings Expression: 100*$No. missings$/$Row count$ Select: Append Selected Columns with the Suffix: _PCT	Input port: Data table from the Upper output port of Node 2 Output port: Statistics table with six rows and 17 columns
Node 4 Column Filter	Column Filter: Include: Column, No. missings, No. missings PCT Exclude all other columns	Input port: Data table from the Output port of Node 3 Output port: Statistics with 6 rows and 3 columns
Node 5 Set Threshold for pct of missing values	Component to allow setting of threshold maximum percent of allowable missing values interactively	Input port: Data table from the Output port of Node 4 Output port: Table with variables not screened out
	Node C1: Component Input	
	Node C2: Double Configuration Enables interactive setting of maximum percent of missing values (min = 0 and max = 100)	
	Node C3: Rule-based Row Filter Expression: $No. missings_PCT$ < $${Double-input}$$ => TRUE	
	Node C4: Transpose	
	Node C5: Component output	
Node 6: Reference Column Filter	This node filters columns from the table at the Upper input port using the table at the Lower input port as reference values Options: Include columns from reference table	Upper input port: Data table from the Output port of Node 1 Lower input port: Data table from the Output port of Node 5 Output port: Table with variable with more than 5% missing filtered out. (1000 rows by 2 columns)

Then, the variances of the normalized variables are computed, and those with variance below a user-defined threshold are filtered out. The workflow in Fig. 5.3 illustrates how to filter by variance. Node descriptions for the workflow are provided in Table 5.4.

The simulated input data contained five variables with variances under 0.05 after 0–1 normalization. These were filtered out, as shown in Table 5.5.

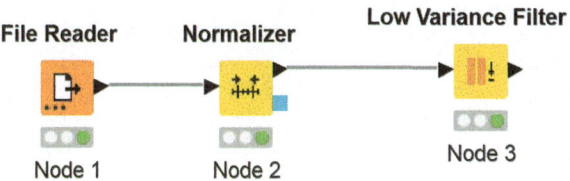

Fig. 5.3 Workflow for screening by variance

Table 5.4 Node descriptions for the Fig. 5.3 workflow

Nodes	Descriptions	Input and output ports
Node 1 File Reader	Read file: SimulatedDataForVarianceScreening.csv	Output port: Data table with 100 rows and seven columns
Node 2 Normalizer	Normalize variables to a range of 0–1 Methods: Include: All numeric variables Settings: Select Min-Max Normalization Min: 0.0 Max: 1.0	Input port: Data table from the Output port of Node 1 Output port: Table with 100 rows and seven columns, normalized to 0–1 range
Node 3 Low Variance Filter	Filters out columns, whose variance is below a user-specified threshold Options: Variance Upper Bound: Range is set to 0.05 with a max variance with range of 0-1 of columns Note: the higher the Variance Upper Bound, more columns will be filtered Include: All numeric variables	Input port: Data table from the Output port of Node 2 Output port: Data table with remaining variables

Table 5.5 Results from filtering simulated data with Variance Upper Bound equal to 0.05

Variable	Variance	Variables after filtering
V1	0.0000	Filtered
V2	0.0099	Filtered
V3	0.2500	V3
V4	0.0949	V4
V5	0.0429	Filtered
V6	0.0429	Filtered
V7	0.0429	Filtered

Selection Using Filters with Target Variables

Another approach to filtering variables is to consider the relationship of each potential predictor with the target variable. Each predictor is assessed one at a time in predicting the target. Four possible situations are shown in Table 5.6 (Brownlee, 2022).

Selection using the approaches listed in Table 5.6 might be helpful in cases with many potential predictors, and the objective is to reduce that number to a more manageable subset. While this is a workable approach, there is no guarantee that the best predictors will be selected.

Table 5.6 Target by predictor combinations

Target variable type	Predictor type	Model	Typical metric
Numerical	Numerical	OLS regression	RMS error
Categorical	Numerical	Logistic regression	Prediction accuracy
Numerical	Categorical	OLS regression	RMS error
Categorical	Categorical	Chi-square	Cramer's V

Selection Using Wrapper Methods
Some algorithms perform feature selection automatically as part of the learning model. Examples include forward, backward, and stepwise, which include predictors that maximize an accuracy criterion. LASSO (least absolute shrinkage and selection operator) models, discussed in the chapters on OLS and logistic regression, also implicitly work to select the best predictors.

5.3 Principal Components Analysis

Another frequently discussed approach to dimension reduction involves forming combinations of predictors into a smaller set and using the smaller set in building a model. The most common technique for this approach is principal component analysis (PCA). Before discussing the details of PCA, two issues make PCA less desirable than feature selection. First, the components formed can be difficult to interpret since they are weighted averages of the predictors, while feature selection maintains the original meaning of the predictors. Second, PCA does not reduce the number of variables that must be collected and stored. When new observations are collected, all the variables used initially in the PCA must be used to create new components.

Principal components analysis (PCA) aims to reduce the dimensionality of a data matrix from p variates to a smaller number of k variates so that most of the information in the original data set is preserved. The smaller number of components is derived in such a way as to capture as much of the actual variation as possible. PCA is used in many disciplines, including psychology, biology, chemistry, education, astronomy, and business, and is one of the oldest multivariate techniques, with early developments by Pearson (1902) and Hotelling (1933).

PCA is a data transformation applied to an $n \times p$ (rows by columns) data. The components are linear functions of the original variables. They are ordered; the first component accounts for the largest variance. Each succeeding component accounts for the largest amount of remaining variance, in turn, subject to being uncorrelated with previously created components.

PCA: An Eigenvector Decomposition
The process of forming components is based on the extraction of eigenvectors and eigenvalues. A data set can be viewed as observations in a multidimensional space. The number of dimensions in the space equals the number of variables in the set. An

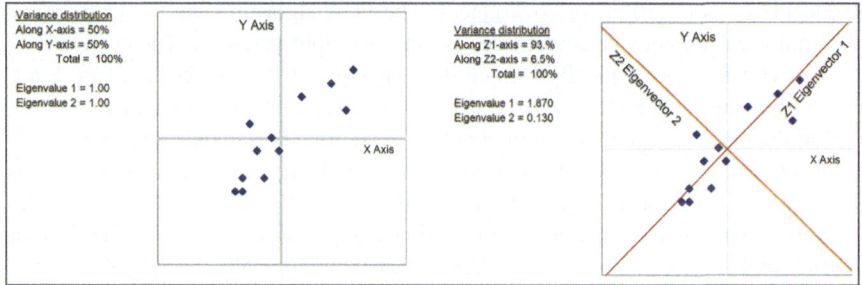

Fig. 5.4 Illustration of eigenvalues

eigenvector is a directed line (or vector) along which the points in a data set can be projected.

Consider the simple case of just two variables. Panel A in Fig. 5.4 shows the plot of 12 observations in two dimensions, X and Y. The variances of both variable X and variable Y are standardized to 1.0, which is typically done in applications of PCA, so the total variance is 2.0. Eigenvalues, which are essentially variances of the extracted components, are also 1.0 for each variable.

The objective of principal components analysis is to rotate the X and Y axes so that the variance along one axis is maximized. This is shown in Panel B of Fig. 5.4. Since there are two variables, there are two eigenvectors. After the first rotated axis is obtained, a second axis is created at right angles to the first, which captures the remaining variance.

The result of the axis rotations is that the first axis captures $100 \times 1.870/(1.8709 + 0.130) = 93.5\%$ of the variance in the small data set. The axis is called eigenvector Z1.

This graphical example explains that PCA is an eigenvector decomposition that finds a series of vectors, each of which maximizes variance along the vector with the restriction that the vectors are uncorrelated. Of course, in a real example, there will be many more dimensions than those used here, but the two-dimensional case allows for a simple geometric illustration. The process for multiple dimensions is the same. The first eigenvector is determined by finding the rotation that maximizes variance (or the eigenvalue). Then, the process is repeated for each additional variable.

Steps in Using PCA

The process of developing principal components involves a series of steps and decisions.

1. Select variables. PCA is designed to be applied to continuous variables. Binary variables can be included if rescaled to 0–1 indicator variables. Other techniques, such as multiple correspondence analysis, should be used with multi-category predictors.

2. Decide on correlations or covariances. The PCA algorithm can be set to either capture correlations or covariances among the input variables. The correlations or covariances are typically formed into a p × p matrix. The results of PCA are not the same for each type of matrix. The correlation matrix standardizes the variables to zero mean and unit standard deviation. This avoids a problem with using the covariance matrix wherein variables with large variances will dominate the result. Also, if a covariance matrix is used and a variable is rescaled, the entire covariance matrix and the resulting components will change. This is an undesirable effect, so the correlation matrix is typically used (Jolliffe and Cadima 2016).

There should be evidence that the correlations or covariances are not so small that PCA makes no sense since PCA relies on relations among variables to form components. If the original variables do not covary, then PCA is not useful.

3. Extract components. Components are formed by creating a set of eigenvectors and corresponding eigenvalues, an eigenvector decomposition. The maximum number of components is the lesser of the number of columns or rows. Still, in most applications, the number of columns is the determinant factor since there are typically many more observations than variables. For dimension reduction, a smaller number of components is retained for further analysis. The overall quality of an approximation of a multidimensional data set is measured by the eigenvalues (or variability) associated with the retained components. So, a decision about the number of components to retain must be made. One rule of thumb is to keep only components for which the associated eigenvalue is greater than or equal to one.

Another approach to selecting the number of components is to plot the eigenvalues by the number of components in a chart called a scree plot, looking for a break where the size of the eigenvalues drops sharply. These two methods may indicate a different number of components to retain. There are no fixed rules, so the judgment of an analyst is required.

4. Compute component scores. Scores are then computed on each component for each observation in the data set. These scores are derived variables, usually fewer in number than the original set, which are uncorrelated and can be used as independent variables in a predictive model.

5.4   Example of Using PCA

A data set with twenty-four characteristics on 205 automobiles was obtained from Kaggle.[4] Eleven continuous variables in the data set were used to demonstrate a principal component analysis (Table 5.7). A workflow for the analysis is shown in Fig. 5.5 and Table 5.9 has the node descriptions.

---

[4] Car information dataset. https://www.kaggle.com/datasets/tawfikelmetwally/automobile-dataset

Table 5.7 Variables in the automobile data set

wheelbase	bore
length	stroke
width	compression-ratio
height	horsepower
curb-weight	peak-rpm
engine-size	

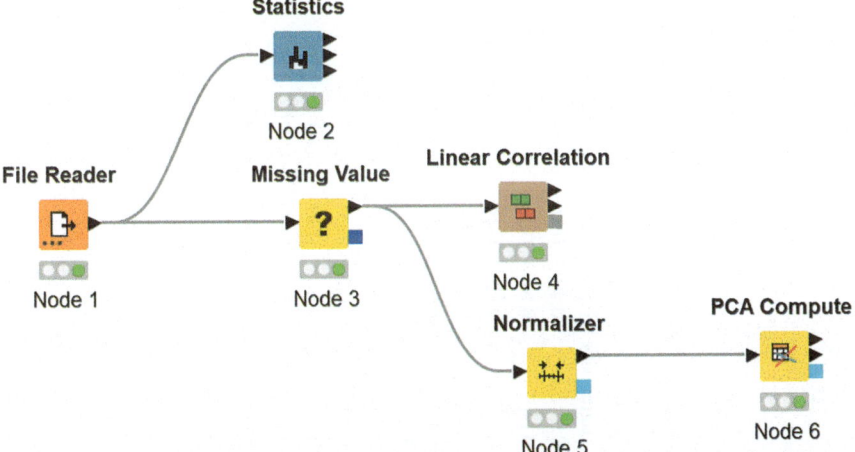

Fig. 5.5 Workflow to impute values, obtain the correlation matrix and extract principal components

Prior to running principal components analysis, the data set was checked for two issues. First, PCA requires a correlation or covariance matrix without missing values, so a KNIME Statistics node was used to check missingness. Four variables had missing entries: horsepower (two missing values), peak-rpm (two missing values), bore (four missing values), and strong (four missing values). The median values for horsepower and peak-rpm were substituted using the Missing Value node so that the cases would not be dropped from the analysis.

Second, an informal check of the existence of relationships among the variables was done by obtaining a correlation matrix which is shown in Table 5.8. There were several reasonably strong correlations among the variables which suggests that principal components analysis can summarize the 11 variables in fewer dimensions.

Since the ranges of the variables differed considerably, the PCA was run on the correlations. The PCA Compute node extracts components from the covariance matrix, but to get around this, the 11 variables were normalized to zero mean and unit standardization. This creates a correlation matrix from the data.

The eigenvalues for the components extracted are shown in Table 5.10. The third column of the table has the cumulative sum of the eigenvalues; note that the sum of all 11 eigenvalues is 11, which is the number of variables since the variables were standardized. The variance of each variable is one since the correlation matrix was

Table 5.8 Correlations among 11 variables in the automobile data set

	wheelbase	length	width	height	curb-weight	engine-size	bore	stroke	compression-ratio	horsepower	peak-rpm
wheelbase	1.00	0.87	0.80	0.59	0.78	0.57	0.49	0.16	0.25	0.35	−0.36
length	0.87	1.00	0.84	0.49	0.88	0.68	0.61	0.13	0.16	0.55	−0.29
width	0.80	0.84	1.00	0.28	0.87	0.74	0.56	0.18	0.18	0.64	−0.22
height	0.59	0.49	0.28	1.00	0.30	0.07	0.17	−0.06	0.26	−0.11	−0.32
curb-weight	0.78	0.88	0.87	0.30	1.00	0.85	0.65	0.17	0.15	0.75	−0.27
engine-size	0.57	0.68	0.74	0.07	0.85	1.00	0.59	0.20	0.03	0.81	−0.24
bore	0.49	0.61	0.56	0.17	0.65	0.59	1.00	−0.06	0.01	0.57	−0.26
stroke	0.16	0.13	0.18	−0.06	0.17	0.20	−0.06	1.00	0.19	0.08	−0.06
compression-ratio	0.25	0.16	0.18	0.26	0.15	0.03	0.01	0.19	1.00	−0.20	−0.44
horsepower	0.35	0.55	0.64	−0.11	0.75	0.81	0.57	0.08	−0.20	1.00	0.13
peak-rpm	−0.36	−0.29	−0.22	−0.32	−0.27	−0.24	−0.26	−0.06	−0.44	0.13	1.00

Table 5.9 Node descriptions for the workflow in Fig. 5.5

Nodes	Descriptions	Input and output ports
Node 1 File Reader	Read File: Automobile_data.csv file	<u>Output port:</u> Table with 205 rows and 11 columns
Node 2 Statistics	Calculate statistics such as minimum, maximum, mean, standard deviation, variance, median, overall sum, number of missing values, row count across all numeric columns, and frequency counts for all continuous and numeric variables values Options: Include all variables	<u>Input port:</u> Data table from the Output port of Node 1 <u>Upper output port:</u> Statistics on continuous variables <u>Middle output port:</u> Histograms for categorical variables (not used in this example) <u>Lower output port:</u> Occurrences (frequencies) for categorical variables

Node 3
Missing value
	Default: imputation:
	    Number (double): Mean
	    Number (integer): Mean
		<u>Input port:</u> Data table from the Output port of Node 1
		<u>Output port:</u> Table with imputed values for missing data (205 rows and 11 columns)

Node 4 Linear Correlation	Options: Include variables listed in Table 5.7	<u>Input port:</u> Data table from the Output port of Node 3

Node 5
Normalizer
	Form a z-score for all variables to extract components from the correlation matrix
	Methods:
	    Include: All numeric files
	    Settings: Z-Score Normalization
		<u>Input port:</u> Data table from the Output port of Node 2
		<u>Output port:</u> Table with 205 rows and 11 columns of normalized data

Node 6
PCA Compute
	Perform principal components analysis
	Settings:
	    Include all predictor variables
	    Exclude: Overall satisfaction
		<u>Input port:</u> Data table from the Output port of Node 5
		<u>Output port:</u> Spectral decomposition (eigenvalues and eigenvectors)

used to form the components. The fourth column of Table 5.10 shows the cumulative percentage of variance extracted by the number of components.

The eigenvalue greater than 1.0 rule of thumb indicates that three components should be retained, accounting for 77% of the variance in the data. As a check, a scree plot showing eigenvalues versus the number of components was created in Excel and is shown in Fig. 5.6. As sometimes happens, the scree plot in this case does not clearly show an "elbow." However, three components seem to be reasonable for demonstration purposes since about 77% of the variation in the original set

Table 5.10 Eigenvalues of the PCA with automobile rating data

Component	Eigenvalues	Cumulative sum	Cumulative PCT
1	5.43	5.43	0.49
2	1.91	7.34	0.67
3	1.15	8.49	0.77
4	0.85	9.34	0.85
5	0.56	9.90	0.90
6	0.41	10.31	0.94
7	0.30	10.61	0.96
8	0.13	10.75	0.98
9	0.11	10.86	0.99
10	0.08	10.94	0.99
11	0.06	11.00	1.00

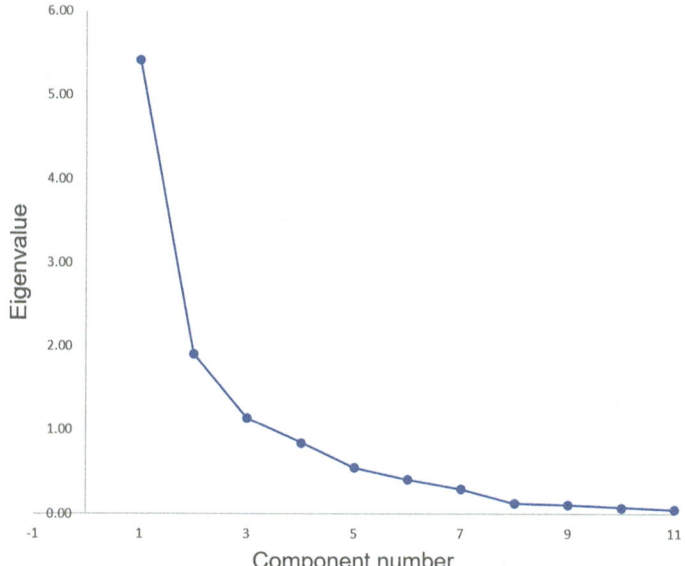

Fig. 5.6 Scree plot of eigenvalues versus the number of components

of variables was captured. This example shows how PCA can be used for dimension reduction. The next section discusses two issues with PCA that analysts should be aware of.

Considerations Regarding PCA

There are two considerations to keep in mind when using PCA as dimension reduction tool. First, to deploy the model with new data, all the original variables must be used to calculate the principal component scores on new data. The dimensionality has been reduced so that subsequent predictive models will have fewer variables,

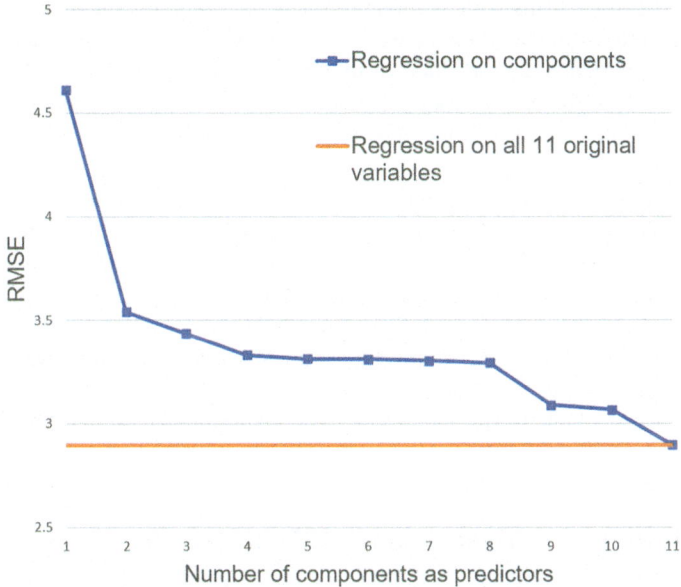

Fig. 5.7 RMS prediction error using from one to 11 components

but measurement and storage of all variables must still be performed. This contrasts with the selection methods described earlier which can drop variables.

The second consideration regarding PCA is that the reduced set of components will nearly always result in diminished predictive accuracy. A regression of highway miles per gallon (MPG) on all 11 predictors was run. (Chap. 6 of this book includes a regression discussion.) The root mean square error (RMSE) using all 11 predictors is 2.898, shown as the horizontal line in Fig. 5.7. Using one component leads to an RMSE of 4.614. Using three components as predictors yields an error of 3.438. The conclusion is that predictive power is lost unless all components are used.

Other Issues with PCA

In some problems, the number of variables p in a data set can exceed the number of observations n. This is true in image analysis, genomics, text analytics, and other contexts. The ordinary PCA model cannot deal with cases where $p > n$. However, Singular Value Decomposition (SVD) can do so quite readily. SVD, used in the KNIME PCA node, can also extract eigenvalues and eigenvectors of data sets where $p > n$, making it a more flexible algorithm.

PCA is sensitive to outliers and errors in the data, and PCA also drops observations that have even a single missing value among the variables. Therefore, it is important to examine the distributions of input variables before PCA and make appropriate adjustments. Values for cases with missing values can be imputed. (Chap. 4 discusses dealing with missing values and outliers.)

As noted earlier, using a PCA model on new data requires measures of all the variables used to form the components. Likely, some of those variables add more to the explained variance. One method of identifying variables to drop is by rotating the components to distinguish more clearly variables with large versus small loadings. The variables with small loadings can be dropped, and PCA repeated on the remaining columns.

Another approach to dimension reduction is to use an adaptation of the LASSO technique. (Discussion of LASSO in the context of regression is discussed in Chap. 6). The LASSO approach has the advantage of forcing some loadings to zero, and variables with zero loading can be dropped. More advanced dimension reduction methods have also been developed (Waggoner 2021).

PCA is an old technique, but many adaptations have been developed to deal with situations where ordinary PCA may not be appropriate. Some adaptations and extensions are available in H2O, R, and Python integrations with KNIME.

5.5 Intuition and Algebra behind Principal Components

A principal component is a linear combination of p variables on n observations. The algebra used to extract principal components in software is beyond the scope of this book and can be found in any text on multivariate analysis. KNIME uses a newer method of extracting components called Singular Value Decomposition. Again, the details are not provided here. Instead, an intuitive example using a tiny data set describes how PCA is done.

For this example, a simple synthetic data set with four observations on three variables was created and is shown in Table 5.11. The data has been standardized to means of zero and standard deviations of 1.0 for each variable. The correlation matrix, which indicates strong correlations among the variables, is shown in Table 5.12.

A component can be created as a linear combination of the three variables for each observation using the following randomly selected weights on each variable: $W1 = 0.510$, $W2 = 0.310$, and $W3 = 0.800$.

Table 5.11 A synthetic data set

Observation	X1	X2	X3
1	−1.432	−1.363	−0.507
2	−0.391	−0.524	−1.183
3	0.651	1.153	1.521
4	1.172	0.734	0.169
Mean	0.000	0.000	0.000
STDev	1.000	1.000	1.000

Table 5.12 Correlations among the variables in Table 5.11

	X1	X2	X3
X1	1.000	0.942	0.594
X2	0.942	1.000	0.797
X3	0.594	0.797	1.000

The sum of the squared weights is:

$$0.510^2 + 0.310^2 + 0.800^2 = 1.00.$$

For observation #1, the component is:

$$0.510(-1.432) + 0.310(-1.363) + 0.800(-0.507) = -1.558.$$

The same linear combination of the three variables can be calculated for all four observations, as shown in Table 5.13. The variance of this first component is 2.174.

This arbitrary component is not useful, however. To compute the first principal component, the weights B1, B2, and B3 are changed to maximize the variance of the component values. A constraint is placed on the weights. Otherwise, maximum variance would be achieved by making the weights arbitrarily large. The constraint is that the sum of the squared values of B1, B2, and B3 = 1.0. The calculations were done in Excel using the Solver add-in.

After the first component is extracted, a second component is calculated again by finding a set of weights that will maximize the variance of that component. The same constraint that the sum of squared values of the weights must be 1.0 is placed, as well as the constraint that the second component is uncorrelated with the first. A third component is then extracted, again with corresponding constraints. The result is that three principal components are extracted such that the first has the most variance, the second has the most remaining variance subject to being uncorrelated with the first, and the third has the most remaining variance subject to being uncorrelated with both the first and second components.

The results are shown in Tables 5.14 and 5.15. Since the original variables were standardized to zero mean and unit standard deviation, the total variance in the transformed variables is 3.00. Note that the three principal components are uncorrelated and that the first component captures more than 85% of the variance in the original data. What has been achieved is that the redundancy in the original data has been used to create components that effectively capture the original variance but transformed to uncorrelated an uncorrelated set of new variables.

Of course, in applied situations, there will be many more variables and observations, and extracting principal components will be accomplished more efficiently than using Excel solver.

Table 5.13 The first component with random weights

Observation	X1	X2	X3	Component 1
1	−1.432	−1.363	−0.507	−1.558
2	−0.391	−0.524	−1.183	−1.308
3	0.651	1.153	1.521	1.906
4	1.172	0.734	0.169	0.960
				Variance = 2.174

Table 5.14 PCA on Table 5.11 synthetic data

Observation	PC1	PC2	PC3	
1	−1.939	0.630	−0.096	
2	−1.181	−0.639	0.152	
3	1.901	0.667	0.091	
4	1.219	−0.658	−0.147	
				Sum of variances
Variance	2.564	0.421	0.015	3.000
% variance	0.855	0.140	0.005	1.000

Table 5.15 Correlations among components

	PC1	PC2	PC3
PC1	1	0	0
PC2	0	1	0
PC3	0	0	1

5.6 Summary

The considerable number of variables available in many projects can be a problem for both the efficiency and effectiveness of predictive analytics. Deciding which variables to include and which to exclude is often a subjective process, but several approaches to dimension reduction are available. Variables can be manually removed based on missing values, near-zero variance, low correlation with the target, or high inter-correlations. Wrapper methods like forward, backward, and stepwise selection can be used to identify a subset of predictors. LASSO can also perform feature selection. Principal components analysis (PCA) creates a new set of features that are linear combinations of the original predictors and fewer in number than the original set. PCA reduces dimensionality but still requires all the original variables for model deployment. In most cases, dimension reduction improves efficiency but risks losing the predictive signal.

References

Brownlee, J. (2022). *How to choose a feature selection method for machine learning.* https://machinelearningmastery.com/feature-selection-with-real-and-categorical-data/#::text=Feature. Accessed 29 July 2023.

Hotelling, H. (1933). Analysis of a complex of statistical variables into principal components. *Journal of Educational Psychology, 24*(6), 417–441.

Hughes, G. F. (1968). On mean accuracy of statistical pattern recognizers. *IEEE Transactions on Information Theory, 14*(1), 55–63.

Jolliffe, I., & Cadima, J. (2016). Principal component analysis: A review and recent developments. *Philosophical Transactions, Royal Society A.* https://doi.org/10.1098/rsta.2015.0202. Accessed 29 July 2023.

Pearson, K. (1902). On lines and planes of closest fit to systems of points in space. *Philosophical Magazine, 2*(11), 559–572.

Silipo, R., Adae, I., Hart, A., & Berthold, M. (2014). *Seven techniques for dimensionality reduction.* https://www.knime.com/sites/default/files/inline-images/knime_seventechniquesdatadimreduction.pdf. Accessed 29 July 2023.

Waggoner, P. D. (2021). *Modern dimension reduction.* Cambridge University Press.

Chapter 6
Ordinary Least Squares Regression

Regression is one of the most used analytics tools. There are several reasons for this. First, it is a logical, linear model with many applications and is conceptually attractive. Linear relationships are easy to think about and perform very well regarding predictive accuracy. Second, the technique itself is easy to program, so there are many, many software tools available. It can be programmed in any language with just a few statements. Third, it is very flexible – it can be applied to many types of problems, even those that initially might not seem to be linear regression candidates. Many phenomena can be cast, at least approximately, into a linear model. Fourth, regression has several different uses. It is often used to build models that explain how one or more independent variables affect a continuous, dependent variable. It is also used for control, in the sense that regression models can help identify cases that are in error or problematic. Finally, it is used for prediction, which is the focus of this chapter.

Regression is used to develop a relationship between a continuous target variable and one or more predictors. The predictors can be continuous numeric, but they need not be. Categorical predictors can be converted to binary numeric variables and used in regression.

Example applications of regression include predicting:

- The price of a house as a function of features such as finished square feet, number of bedrooms, and size of the lot.
- The price of used cars given age, mileage, options, and make, among other characteristics.
- A patient's blood pressure as a function of various drug dosages and the patient's age, sex, and weight.
- Crop yield as a function of the type of fertilizer, rainfall, and soil type.
- The number of people that might access a particular website.
- The performance of professional athletes as a function of different training regimens.

F. Acito, *Predictive Analytics with KNIME*,
https://doi.org/10.1007/978-3-031-45630-5_6

6.1   Intuition About Simple Linear Regression

Simple linear regression models consist of a single predictor, which might be continuous or categorical, and a continuous dependent variable. For example, consider the 11 observations shown in Table 6.1.

The regression equation is developed to predict y_i as accurately as possible from the x_i values. The model, shown in Eq. (6.1), shows that y_i is a function of a constant (β_0) plus a coefficient (β_1) times x_i plus an error term (ε_i).

$$y_i = \beta_0 + \beta_1 \times x_i + \varepsilon_i \tag{6.1}$$

The estimates $\hat{\beta}_0$ and $\hat{\beta}_1$, denoted by "hats," are found by minimizing the sum of the squared errors (*SSE*) between y_i and the estimates $\hat{y}_i$. Equation (6.2) shows the minimization problem:

$$Find\ \hat{\beta}_0\ and\ \hat{\beta}_1\ to\ minimize\ SSE = \sum_{i=1}^{n}\left(y_i - \hat{y}_i\right)^2 \tag{6.2}$$

The solutions to the minimization problem are found with calculus and are shown in Eqs. (6.3) and (6.4).

$$\hat{\beta}_1 = \frac{\sum\left(x_i - \bar{x}\right)\left(y_i - \bar{y}\right)}{\sum\left(x_i - \bar{x}\right)^2} \tag{6.3}$$

$$\hat{\beta}_0 = \bar{y} - \hat{\beta}_1 \times \bar{y} \tag{6.4}$$

where $\bar{x}$ and $\bar{y}$ are the means of x and y, respectively.

Figure 6.1 illustrates the concept of least squares. The vertical lines from each observation to the regression line represent the error. Note that the error can be

Table 6.1 Observations in example data set

Observation	x	y
1	6.1	27
2	8.7	44
3	7.8	35
4	9.8	42
5	5.5	24
6	2.2	5
7	3.5	23
8	3.9	16
9	4.0	15
10	9.1	31
11	6.9	14

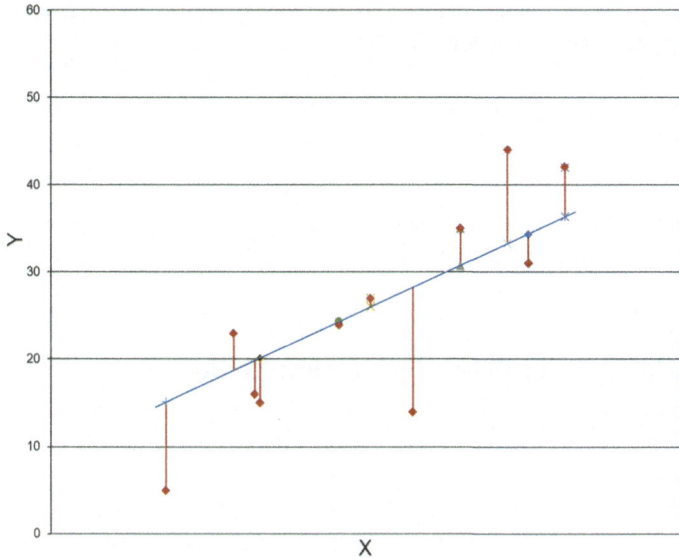

Fig. 6.1 Illustration of the least squares principle

positive or negative; the error is squared so that both positive and negative differences contribute to the total error. The total error can be reduced by moving the intercept and slope of the line. The objective is to make the sum of the squared errors as small as possible.

The more general case of multiple regression is discussed in the next section.

6.2 Multiple Regression

In many situations, there are several predictor variables. Running separate simple regressions on each of several predictors will not yield the desired results (unless all the predictors are completely uncorrelated.)

The multiple regression model is given in Eq. (6.5). The model shows that y_i is a function of a constant (β_0) plus several coefficients (β_j) times each of the predictors, x_{ij}, plus an error term (ε_i) where i refers to the observation number.

There can be any number of predictors if the number of observations, n, is much greater than the number of predictors, p. The betas are the regression coefficients, indicating how a change in a predictor affects the target variable Y. For example, if x_1 increases by 10, Y will increase by β_1 times 10.

$$y_i = \beta_0 + \beta_1 x_{i1} + \cdots + \beta_j x_{ij} + \cdots \beta_p x_{ip} + \varepsilon_i \qquad (6.5)$$

The coefficients shown in the model in (6.5) are not observed but must be estimated. The estimated model is shown in Eq. (6.6).

$$\hat{y}_i = \hat{\beta}_0 + \hat{\beta}_1 x_{1j} + \cdots + \hat{\beta}_j x_{ji} + \cdots \hat{\beta}_p x_{ip} \tag{6.6}$$

Once again, the $\hat{\beta}_j$ coefficients are found by minimizing the sum of squared errors-between the actual and predicted values, or $SSE = \sum_{i=1}^{n} (y_i - \hat{y}_i)^2$. The coefficients are derived using calculus and algebra.

6.3 Building a Predictive Regression Model

Several issues need to be considered when building a predictive regression model. The number of observations must be greater than the number of predictor variables. While textbooks on regression typically say that 5–10 times the number of predictors is the minimum number of observations, in data mining, where several models will be tried, at least 100 to 1000 observations per predictor variable is recommended. Another reason for suggesting such a high ratio of observations to predictors is that the data is typically divided into subsets for training, testing and validation.

Variable Types in Regression

It is helpful to consider the types of variables encountered in regression applications and, indeed, in any analytics project. Just because a data set contains numbers, numbers may be treated differently when doing analyses depending on what measurement level the numbers represent.

A basic framework to classify variables based on measurement properties uses three categories: nominal (or categorical), ordinal, and continuous (sometimes called quantitative). Nominal variables represent categories without any intrinsic ordering. With ordinal variables, adjacent levels indicate increasing or decreased values. Continuous variables include the features of nominal and ordinal measures plus the feature of equal increments between adjacent values.

Ordinary regression is assumed to have a continuous target variable. Predictor variables in regression can be nominal, ordinal, or continuous in any combination. However, nominal and ordinal variables must be coded to be meaningfully included in the regression.

Coding Nominal Variables

Examples of nominal variables include color ("red," "green," and "blue"), location ("city," "suburb," or "rural"), gender, country, and customer type ("individual" or "business"). Numbers may be used to represent nominal classifications (e.g., 1 = "red", 2 = "green", and 3 = "blue"), but the numbers do not indicate quantities and should not be used without coding.

Nominal variables, which have $k \geq 2$ levels, can be coded using indicator variables (aka "dummy variables") for regression. Because the regression model usually includes an intercept term, for a k-level nominal variable, one level is omitted from the coding scheme. So, only k-1 indicator variables are required, and it does not matter which level is omitted. If k indicators are created and submitted to most regression programs, an error is flagged because one of the indicators is redundant.

For example, a regression might include political party as a predictor. Since this is a nominal variable, it is necessary to code political parties using binary indicator variables. Table 6.2 shows that the four category levels are coded using three indicator variables. The regression predictors would include Party1, Party2, and Party3, but the original variable, political party, would be dropped from the analysis. Notice that the level "Republican" is coded with all three zero values. The three-party indicators use Republican as a reference value.

When used as predictors in regression with a continuous target variable y, the following interpretations can be made for the three indicator variables:

- The coefficient on Party1 estimates the difference in y between Democrats and Republicans.
- The coefficient on Party2 estimates the difference in y between Independents and Republicans.
- The coefficient on Party3 estimates the difference in y between Other and Republicans.
- The difference in y due to Democrat versus Independent would be estimated by the difference between the coefficient on Party1 minus the coefficient on Party2.
- The difference between the coefficient on Party2 minus the coefficient on Party3 would estimate the difference in y due to Independent versus Other.
- The difference between the coefficient on Party1 minus the coefficient on Party3 would estimate the difference in y due to Democrat versus Other.

Table 6.2 Example of coding a nominal variable

Political party	Indicator (dummy) variables		
	Party1	Party2	Party3
Republican	0	0	0
Democrat	1	0	0
Independent	0	1	0
Other	0	0	1

Table 6.3 Example of coding an ordinal variable

	Indicator (dummy) variables		
Temperature	Temp1	Temp2	Temp3
Cool	0	0	0
Warm	1	0	0
Hot	1	1	0
Very hot	1	1	1

Coding Ordinal Variables

Ordinal variables imply direction, ordering, or relative magnitude. However, unlike continuous variables, the increments between levels are not necessarily equal. Examples of ordinal variables include:

- Clothing size (small, medium, or large).
- Agreement level (Yes, Maybe, No).
- Income category (Above average, Average, Less than average).

For example, consider an ordinal temperature variable containing four levels: Cool, Warm, Hot, and Very hot. Indicator coding can also be used for this example. As with nominal coding, the number of indicator variables is one fewer than the number of levels. The coding scheme (shown in Table 6.3) is set to capture the ordinal property (Lyons, 1971). Again, Temp1, Temp2, and Temp3 would be used as predictors, but Temperature would not.

    When used as predictors in regression with a continuous target variable y, the following interpretations can be made for the three indicator variables:

- The difference in y between Warm and Cool would be estimated by the coefficient on Temp1.
- The difference in y between Hot and Cool would be estimated by the coefficient on Temp1 plus the coefficient on Temp2.
- The difference in y between Very Hot and Cool would be estimated by the coefficient on Temp1 plus the coefficient on Temp2 plus the coefficient on Temp3.
- The difference in y between Very Hot and Hot would be estimated by the coefficient on Temp3.
- The difference in y between Very Hot and Warm would be estimated by the coefficient on Temp2 plus the coefficient on Temp3.

6.4   Nonlinear Relationships

The OLS model assumes that the relationship between continuous predictors (individually and in combinations) and the y target variable is linear. If the nonlinearity is slight, the regression model can still be a useful approximation of the relationship.

In some cases, however, strongly nonlinear relationships can seriously distort the regression estimates, and the model will not make accurate predictions.

Detecting possible nonlinearity with a single predictor can be seen by plotting the target variable *y* versus the predictor. For multiple regression, however, plotting the target variable versus each predictor can be misleading. Instead, with multiple regression, one approach is to form a set of residuals by regressing the target on all but one held-out predictor at a time. The residuals thus created can be plotted versus the held-out predictor. This will show the relationship with a given predictor controlling for the other predictors.

Regression with a Nonlinear Relationship

Figure 6.2 shows a plot of simulated data where the *y-values* are related to the *x-values* in a nonlinear fashion.

After the regression of *y* on *x*, predicted values are obtained and plotted as shown in Fig. 6.3. The linear model in the dashed line fits reasonably well, but the nonlinearity produces negative errors at low and high values of *x* and positive errors at mid-values of *x*.

The effects of nonlinearity can be seen more clearly by plotting the residual values $y_i - \hat{y}_i$ versus *x* in Fig. 6.4. Residual plots make the nonlinearity much more apparent.

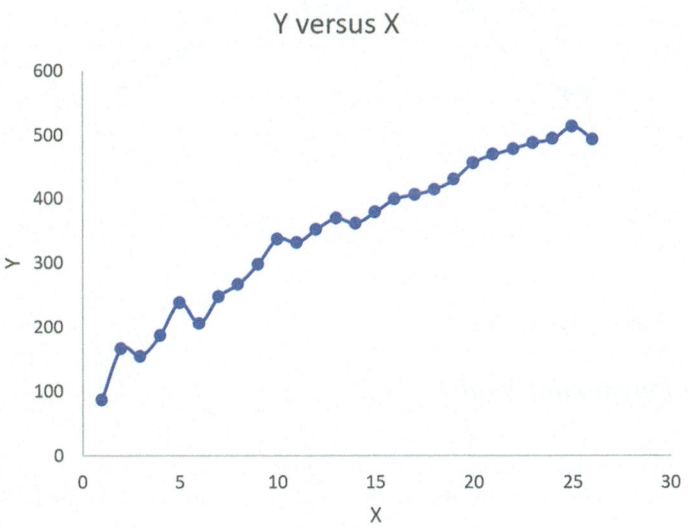

Fig. 6.2 Illustration of a nonlinear relationship

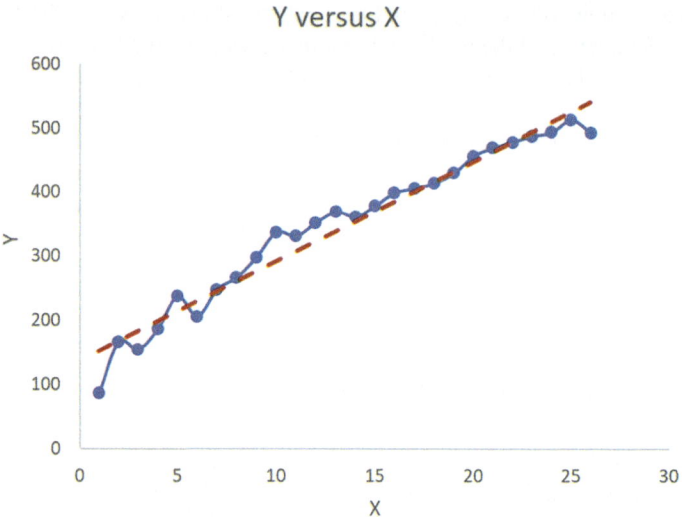

Fig. 6.3 Linear regression of y versus x

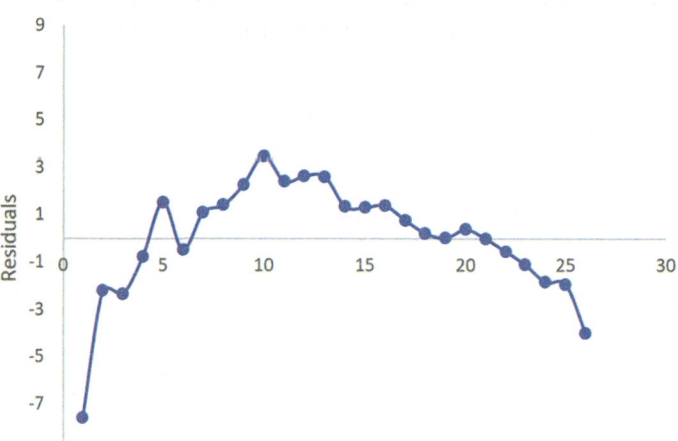

Fig. 6.4 Residuals versus X values

Using a Polynomial Model

A polynomial model, which includes the square of each x_i was run, and y_i was regressed on x_i plus x_i^2. Note that both x_i and x_i^2 are included as predictors. The model shown by the dashed line in Fig. 6.5 captures the nonlinear relationship.

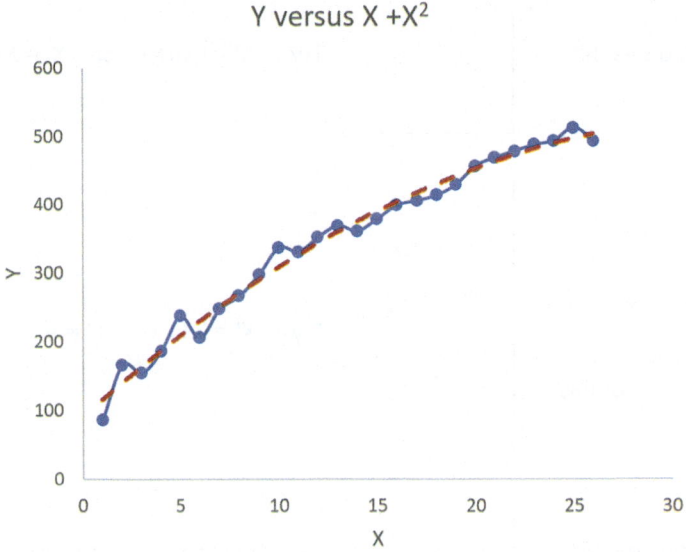

Fig. 6.5 Polynomial regression of y versus x

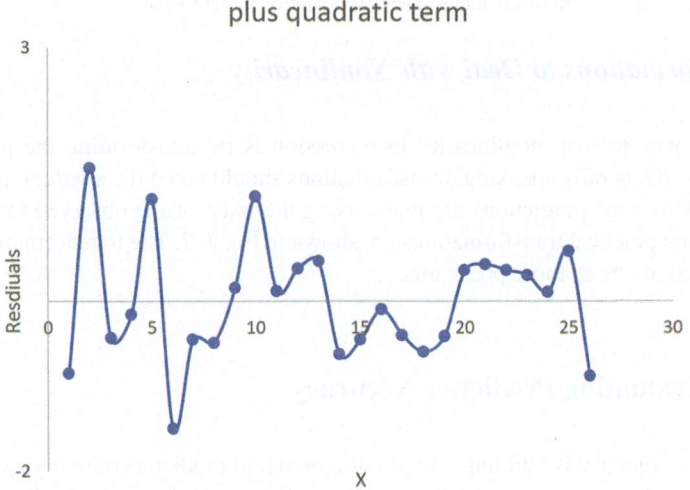

Fig. 6.6 Residual plot with polynomial regression

The residuals from the polynomial model were computed and plotted versus *x* in Fig. 6.6. Note that the range of the residuals is less than 4 in contrast to the range in Fig. 6.4 which indicates a range of about 11.

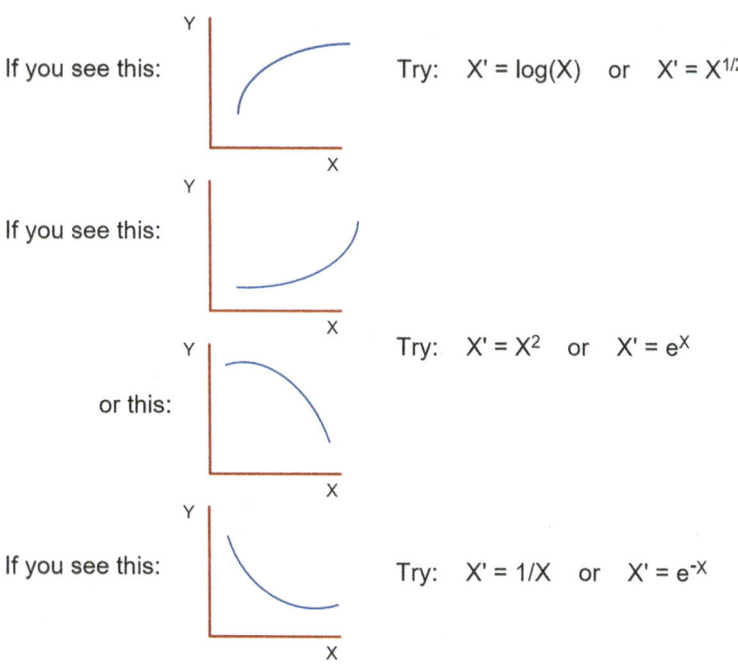

Fig. 6.7 Examples of transformations to achieve linearity in regression

Transformations to Deal with Nonlinearity

Another way to treat nonlinearity in regression is by transforming the predictor variables. (Generally speaking, transformations should be on the *x values*, not the *y values*. This way, predictions are made using the scale of the observed target values.) Some practical transformations are shown in Fig. 6.7. The transformations can be applied to one or more predictors.

6.5   Evaluating Predictive Accuracy

When the objective is building a regression model to predict a continuous variable, several metrics can assess predictive accuracy. R^2, one of the most widely available statistics in statistical software, is not considered a useful evaluation measure of predictive accuracy. (Instead, R^2 is most useful when an exploratory regression is a model created to show the model's fit to the data, which is not necessarily identical to predictive accuracy.)

Three commonly used statistics to assess prediction accuracy with test or hold-out data are mean absolute error (MAE), root mean square error (RMSE), and mean absolute percentage error (MAPE). The calculation of each of these measures is based on the prediction error with the hold-out data, $e_i = y_i - \hat{y}_i$. The errors, e_i can be either positive or negative, so as the total prediction errors are computed, the signs must be ignored. This can be done by squaring the e_i or taking the absolute value. The formulas for the three measures are given below in Eqs. (6.6), (6.7), and (6.8). These metrics are provided in the Numeric Scorer node in KNIME.

$$Mean\,absolute\,error = MAE = \frac{\sum_{i=1}^{n}|e_i|}{n} \tag{6.7}$$

$$Root\,mean\,square\,error = RMSE\sqrt{\frac{\sum_{i=1}^{n}e^2}{n}} \tag{6.8}$$

$$Mean\,absolute\,percent\,error = MAPE = \frac{\sum_{i=1}^{n}\left|\frac{e_i}{y_i}\right|}{n} \tag{6.9}$$

Note that MAPE does not work if any $y_i = 0$

6.6   Example Applications of Regression

This section will demonstrate multiple regression using the analytics tools available in KNIME, including ordinary least squares, stepwise regression, and regularized regression models.

Predicting Home Prices

This example uses a subset of the well-known Ames, Iowa, housing price data published in 2011 (De Cock, 2011). The analysis estimates the selling prices of homes as a function of characteristics such as quality, living area, and year constructed. A subset of the predictors from the complete Ames data set was selected for this example. The predictor variables for this example are listed in Table 6.4, along with the target variable, selling price. The data set contains 2930 observations and 20 predictor variables, 11 continuous and nine categorical. The levels of some of the categorical variables were grouped to avoid sparseness.

The workflow for analyzing the housing price data is in Fig. 6.8, and the node descriptions are in Table 6.5). (Note that dummy variables are automaticallycreated for categorical predictors in the KNIME regression node.)

Table 6.4 Variables in the Ames data set

Variable	Description	Type of measure
Sale_price	Sale price in US dollars	Continuous
MS_Zoning	Zoning classification	Categorical with 7 levels; collapsed to 5
Living_Area	Above grade living area square feet	Continuous
Lot_Area	Area of lot in square feet	Continuous
Neighborhood	Physical locations within Ames city limits	Categorical with 28 levels; collapsed to 23
Proximity	Proximity to various conditions	Categorical with 8 levels; collapses to 5
Bldg_Type	Type of dwelling	Categorical with 5 levels
House_Style	Style of dwelling	Categorical with 8 levels; collapses to 3
Overall_Qual	Rating of the overall material and finish	Continuous scale: 1 to 10
Overall_Cond	Rating of the overall condition of the house	Continuous scale: 1 to 10
Exter_Qual	Evaluation of the exterior material quality	Categorical with 4 levels
Kitchen_Qual	Evaluation of the kitchen quality	Categorical with 5 levels; collapses to 3
Year_Built	Year constructed	Continuous
TotRms_AbvGrd	Total number of rooms above grade	Continuous
Bedroom_AbvGr	Number of bedrooms above grade	Continuous
Full_Bath	Number of full baths	Continuous
Half_Bath	Number of half baths	Continuous
Garage_Cars	Size garage (no. of cars)	Categorical with 4 levels
Central_Air	Central air conditioning	Categorical with 2 levels
Fireplaces	Number of fireplaces	Continuous
Yr_Sold	Year of sale	Continuous

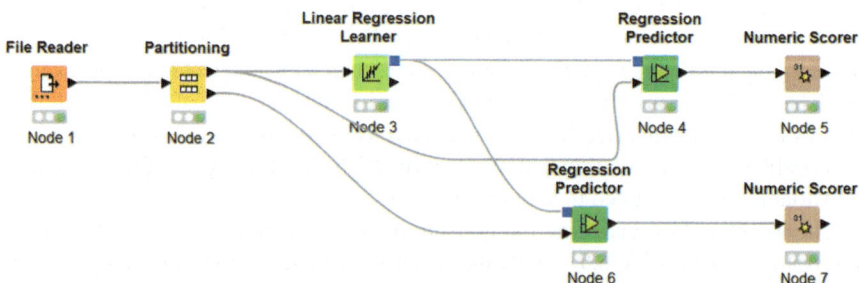

Fig. 6.8 Workflow for OLS with Ames data

Table 6.5   Node descriptions for the Fig. 6.8 workflow

Nodes	Descriptions	Input and Output ports
Node 1 File reader	Settings: File: SubsetAmesHousing.csv	<u>Output port:</u> data table with 2930 rows and 21 columns
Node 2 Partitioning	First partition is the training set relative (%): 70%. Draw randomly Use random seed: 123	<u>Input port:</u> data table from the output port of node 1 <u>Upper output port:</u> training data (first partition); 2051 rows and 21 columns <u>Lower output port:</u> test data (test data); 879 rows and 21 columns

Node 3
Linear
regression
learner
	OLS on training data
	Settings:
	    Target: Sales_price
	    Predictors: all other variables
		<u>Input port:</u> data table from the upper output port of node 2
		<u>Upper output port:</u> OLS model
		<u>Lower output port:</u> coefficients and statistics (not used here)

Node 4 Regression predictor	Predicts Sales_price using a regression model with the training data. A new column to the input table containing the prediction for each row	<u>Upper input port:</u> model from the upper output port of node 3 <u>Lower input port:</u> data table from the upper output port of node 2 <u>Output port:</u> training data plus predicted Sales_price

Node 5
Numeric
scorer
	Score the training data
	Options:
	    Reference column: Sales_price
	    Predicted column: prediction (Sales_price)
		<u>Input port:</u> data table from the output port of node 4
		<u>Output port:</u> performance statistics on training data

Node 6
Regression
predictor
	Predicts Sales_price on test data
	Options:
	    Reference column: Sales_price
	    Predicted column: prediction (Sales_price)
		<u>Upper input port:</u> data table from the output port of node 4
		<u>Lower input port:</u> data table from the lower output port of node 2
		<u>Output port:</u> test data plus predicted Sales_price

Node 7
Numeric
scorer
	Score test data
	Options:
	    Reference column: Sales_price
	    Predicted column: prediction (Sales_price)
		<u>Input port:</u> data table from the output port of node 6
		<u>Output port:</u> performance statistics on test data

The results for the training and test sets were comparable, with the accuracy measures on the test set slightly better (Table 6.6). (The predictive accuracy can be higher with the test data due to the random sampling process that creates the training and test subsets.) The regression model had 67 predictors, several of which were not statistically significant. Therefore, a forward stepwise OLS mode was tried to determine if comparable accuracy could be obtained with fewer predictors.

Table 6.6 RMSE results for
OLS regression

| | | Training set | Test set |
Model	RMSE	RMSE
OLS	28,770.8	28,260.7

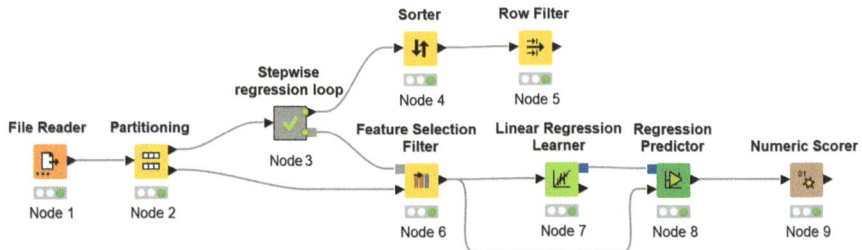

Fig. 6.9 Workflow for stepwise regression with Ames data

Comparison of OLS and Stepwise Regression

KNIME does not have a pre-built forward stepwise regression node, but the Feature
Selection Loop can be used for stepwise. The workflow shown in Fig. 6.9 was constructed to run the model. Descriptions of each node are in Table 6.7. The basic
logic is that the workflow will loop through all the predictors and select the one that
minimizes the prediction error. Since the data set relies on the test set to select predictors, a separate validation set was created to ensure the model is not overfitting.

Table 6.8 shows the RMSE for the stepwise model as well as for the OLS model.
The final accuracy was measured on the test set for the stepwise model and the test
set for the OLS model. The stepwise model had slightly better prediction accuracy
on the test set compared with the OLS model.

Using Regularization for Predictor Selection

Regularization can reduce the complexity of regression models, improve generalizability, and speed up model learning without a large reduction in accuracy.
Regularization works by modifying the model by adding a penalty function. The
penalty function increases with the size of estimated beta coefficients, so the overall
effect is to develop a model with smaller coefficients. Three types of penalty functions will be discussed here. The penalty with L1 regularization (also known as
LASSO or Least Absolute Shrinkage and Selection Operator) (Tibshirani, 1966) is
a function of the sum of the absolute values of the beta coefficients. L2 regularization, also known as Ridge Regression (Hoerl & Kennard, 1970), uses a penalty

Table 6.7 Node descriptions for the Fig. 6.9 workflow

Nodes	Descriptions	Input and output ports
Node 1 File reader	Settings: File: SubsetAmesHousing.csv.	<u>Output port:</u> data table with 2930 rows and 21 columns
Node 2 Partitioning	First partition is the training set relative (%): 70 Draw randomly Use random seed: 123	<u>Input port:</u> data table from the output port of node 1 <u>Upper output port:</u> training and test data (first partition); 2051 rows and 21 columns <u>Lower output port:</u> validation data; 879 rows and 21 columns
Node 3 Stepwise regression loop	Metanode to evaluate each predictor in a stepwise regression	<u>Input port:</u> data table from the upper output port of node 2 <u>Upper output port:</u> result table with the number of features, root mean square prediction error, and removed feature <u>Lower output port:</u> feature selection model
	Node M3_1 feature selection loop start Node M3_2 partitioning Node M3_3 linear regression learner Node M3_4 regression predictor Node M3_5 numeric scorer Node M3_6 feature selection loop end	
Node 4 Sorter	Sorting filter Sort by: root mean squared error Check: ascending	<u>Input port:</u> data table from the upper output port of node 3 <u>Output port:</u> result table with number of features, root mean square prediction error, and removed feature, sorted by root mean square error, from lowest to highest
Node 5 Row filter	Retain row with lowest root mean square error Filter criteria: Check: include rows by number Row number range: 1 to 1	<u>Input port:</u> data table from the output port of node 4 <u>Output port:</u> filtered row
Node 6 Feature selection filter	This node selects the subset of columns identified in the feature selection Loop and applies the subset selection to the validation data set Column selection: Check: include static columns (to include the target Sales_price) Select: select features with best score	<u>Upper input port:</u> lower model port from node 3 <u>Lower input port:</u> the lower output port of Node 2 (the validation data) <u>Output port:</u> the input data set with some columns filtered out

(continued)

Table 6.7 (continued)

Nodes	Descriptions	Input and output ports
Node 7 Linear regression learner	OLS on the validation data Settings: Target: Sales_price Predictors: variables selected in feature selection filter	Input port: data table from the output port of node 6 Upper output port: OLS model with filtered predictor variables Lower output port: coefficients and statistics
Node 8 Regression predictor	Predicts Sales_price using a regression model with the validation data using the columns from the feature selection loop A new column to the input table containing the prediction for each row	Upper input port: OLS model from upper output port of node 7 Lower input port: data table from the output port of node 6 Output port: predictions on the validation data from node 1
Node 9 Numeric scorer	Score the validation data Options: Reference column: Sales_price Predicted column: prediction (Sales_price)	Input port: data table from the output port of node 8 Output port: performance statistics on the validation data

Table 6.8 RMSE results for OLS and stepwise regression

Model	Number of predictors	Test set RMSE	Validation set RMSE
OLS	59	28,770.8	28,260.7
Stepwise OLS	51	30,995.7	27,418.0

function equal to the sum of the squared beta coefficients. The third type of regularization, called Elastic Net regression (Zou & Hastie, 2005), combines L1 and L2 regularization.

Regularization Formula

L1, L2, and Elastic Net regularization are captured in the following formula (6.10), which adds a penalty to the total sum of errors squared:

$$SSE = \sum_{i=1}^{n}\left(y_i - \hat{y}_i\right)^2 + \lambda\left[\left(1-\alpha\right)^2 \times \beta_i^2 + \alpha \times |\beta_i|\right] \tag{6.10}$$

The λ parameter controls the amount of regularization, and α controls the distribution between L1 and L2 penalties. If $\lambda > 0$, and $\alpha = 1$, L1 regularization is applied; if $\lambda > 0$, and $\alpha = 0$, L2 regularization is used; if $\lambda > 0$, and $0 < \alpha < 1$, elastic net regularization is applied; if lambda = 0, no regularization is applied; alpha is ignored.

L1 regularization forces the optimization procedure to assign smaller values to the coefficients in the regression model. The lambda parameter controls how much emphasis is given to the penalty term. As lambda increases, the coefficients will be reduced, with some even forced to be zeros. A model with just the most important predictors is created with some coefficients forced to zero. (L2 regularization causes the optimization procedure to assign smaller values to the coefficients in the model, but the coefficients are never pushed to zero.) Elastic Net does both: it forces some coefficients to zero and reduces the magnitude of others.

By reducing the magnitude of the coefficients, no single predictor is allowed to have a dominant impact on the target value. Even if there are unusual cases in the training data with large or small values on some predictors, the predicted target values will not change dramatically. Hence, the model becomes more stable and generalizable to new data.

The following KNIME workflow uses L1 regularization with the Ames data. L1 regularization, available in KNIME with the H2O Machine Learning tools, was selected to demonstrate the simplification possible by setting unneeded coefficients to zero. The workflow diagram is in Fig. 6.10.

The node descriptions are provided in Table 6.9. The lambda value in the L1 regularization was set by trial and error to achieve balanced accuracy for the training and test data sets.

Comparisons of Prediction Accuracy

The prediction accuracy statistics comparing the OLS, stepwise, and regularized regression models are shown in Table 6.10. The average of the RMSE values on the test sets for the three models is shown in the table; the three models differed from the average by about plus or minus 5%. Both the stepwise and L1 regularization models reduced the number of predictors. The stepwise model required a more complex workflow since a feature selection loop was needed. The L1 regularization model is an attractive approach to variable selection.

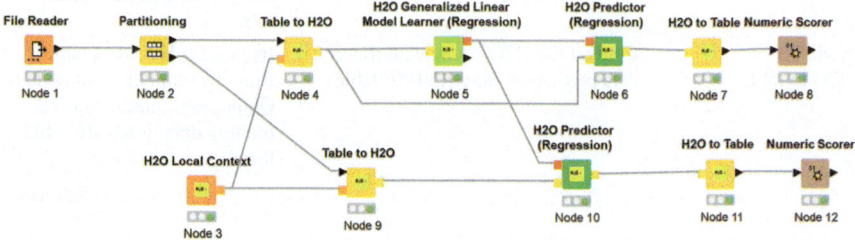

Fig. 6.10 Workflow for L1 regularized regression on Ames data

Table 6.9 Node descriptions for the Fig. 6.10 workflow

Nodes	Description	Input and output ports
Node 1 File reader	Settings: file: SubsetAmesHousing.csv.	Output port: data table with 2930 rows and 21 columns
Node 2 Partitioning	First partition is training setRelative (%): 70% Draw randomly Use random seed: 123	Input port: data table from the output port of node 1 Upper output port: training data (first partition); 2051 rows and 21 columns Lower output port: test data; 879 rows and 21 columns
Node 3 H2O local context	Create a locally running H2O instance to enable the H2O tools for use in KNIME	Output port: local H2O context
Node 4 Table to H2O	Convert the training data from a KNIME table to H2O data frame Column selection: include all variables	Upper input port: data table from the upper output port of node 2 Lower input port: the local H2O context from node 3 Output port: training data converted to H2O table

Node 5
H2O generalized linear model learner (regression)
	This node is used to create generalized linear model, allowing regularized regression
	General settings:
	    Target column: MEDV
	    Include: all predictors
	    Algorithm settings:
	    Family: Gaussian
	        Link: family default
	        Set alpha: 1 to produce L1 regularization
	            Set lambda: 210 to produce L1 regularization
		Input port: H2O data table from output port of node 4
		Upper output port: H2O GLM regression model
		Lower output port: not used here

Node 6 H2O predictor (regression)	Predictions using training data and H2O GLM model	Model input port: H2O GLM regression model from node 5 Lower input port: training data converted to H2O table Output port: predictions on training data in H2O table form
Node 7 H2O to table	Convert the H2O frame with training data predictions to a KNIME table	Input port: H2O data table from the output port of node 6 Output port: predictions on training data in KNME table form

(continued)

Table 6.9 (continued)

Nodes	Description	Input and output ports

Node 8
Numeric scorer
	Score training data
	Options:
	   Reference column: Sales_price
	   Predicted column: prediction
	   (Sales_price)
		Input port: data table from the output port of node 7
		Output port: performance statistics on training data

Node 9 Table to H2O	Convert the test data from a KNIME table to H2O data frame Column selection: Include all variables	Upper input port: data table from the lower output port of node 2 Lower input port: H2O context Output port: data converted to H2O table
Node 10 H2O predictor (regression)	Predictions using test data and H2O GLM model	Upper input port: H2O GLM regression model from node 5 Lower input port: test data converted to H2O table Output port: predictions on test data in H2O table form
Node 11 H2O to table	Convert the H2O frame with test data predictions to a KNIME table	Input port: H2O data table from the output port of node 10 Output port: predictions on test data in table form

Node 12
Numeric scorer
	Score test data
	Options:
	   Reference column: Sales_price
	   Predicted column: prediction
	   (Sales_price)
		Input port: data table from the output port of node 11
		Output port: performance statistics on test data

Table 6.10 RMSE Accuracy results for OLS, stepwise, and L1 regularization

Model	Number of predictors	Training set RMSE	Test set RMSE
OLS	59	28,770.80	28,260.70
Stepwise OLS	51	30,995.70	27,418.00
L1 regularized	37	29,371.00	28,525.20
Average		29,629.20	28,277.80

6.7 Summary

Ordinary least squares regression models the relationship between a continuous target variable and one or more predictor variables, which can be continuous or categorical. The regression equation models the target variable as a function of a constant term, coefficients on each predictor variable, and an error term. Estimates of the constant term and the coefficients are derived by minimizing the sum of squared errors between the actual and predicted target values.

When building a regression model, issues include having enough observations relative to predictors, checking for nonlinear relationships, and evaluating predictive accuracy on a test set using metrics such as RMSE Root Mean Square Error (RMSE), Mean Absolute Error (MAE), and Mean Absolute Percent Error (MAPE). Variable selection methods like stepwise regression and regularization (L1, L2, elastic net) can help reduce model complexity and improve generalizability.

The Ames housing price data was analyzed using ordinary regression, stepwise regression, and LASSO. As measured by Root Mean Square Error (RMSE), prediction accuracy on the test data was comparable among the three models. However, the stepwise and LASSO models simplified the model by reducing the number of predictors.

References

De Cock, D. (2011). Ames, Iowa: Alternative to the Boston Housing Data as an end of semester regression project. *Journal of Statistics Education, 19*(3) https://jse.amstat.org/v19n3/decock.pdf

Hoerl, A., & Kennard, R. W. (1970). Ridge regression: Biased estimation for nonorthogonal problems. *Technometrics, 12*(1), 55–67.

Lyons, M. (1971). Techniques for using ordinal measures in regression and path analysis. *Sociological Methodology, 3*, 147–171.

Tibshirani, R. (1966). Regression shrinkage and selection via the lasso. *Journal of the Royal Statistical Society Series B (Methodological), 58*(1), 301–320.

Zou, H., & Hastie, T. (2005). Regularization and variable selection via the elastic net. *Journal of the Royal Statistical Society, Series B (Methodological), 67*(Part 2), 301–320.

Chapter 7
Logistic Regression

Many analytics projects are designed to predict which of two outcomes is more likely. The target variable in such applications is binary with two values usually expressed as "1" or "0." The possible values are typically called classes or levels. Examples of binary outcomes include:

- Buy/Not buy.
- Success/Failure.
- Heart disease/No heart disease.
- Continue/Not continue.
- Fraud/No fraud.
- Cancer/No cancer.
- Vote yes/Vote no.

Of course, this is just a partial list since binary outcomes are of concern in many areas, including human resources, marketing, medicine, education, political science, and criminology.

While ordinary linear regression (OLS) is appropriate when the objective is to predict a continuous numerical target, there are better tools to use when the target variable only takes on discreet values. Figure 7.1 compares the linear predictions using ordinary regression with a single predictor, x, with the binary predictions from logistic regression. The target variable, y, is assumed to be binary. But OLS will produce unbounded estimates in the positive and negative directions. Logistic regression, in contrast, makes predictions ranging from 0.0 to 1.0, which can be interpreted as the probability that the target variable takes on the value of 1 versus 0. Visually, the logistic transformation "bends" the two ends of the straight line of linear regression to the allowed range.

This chapter begins with an intuitive explanation of logistic regression and then introduces the logistic function. Analyses with binary outcomes are discussed in the first part of this chapter. Logistic regression can also be applied to problems with multi-class outcomes, which will be demonstrated.

© The Author(s), under exclusive license to Springer Nature Switzerland AG 2023
F. Acito, *Predictive Analytics with KNIME*,
https://doi.org/10.1007/978-3-031-45630-5_7

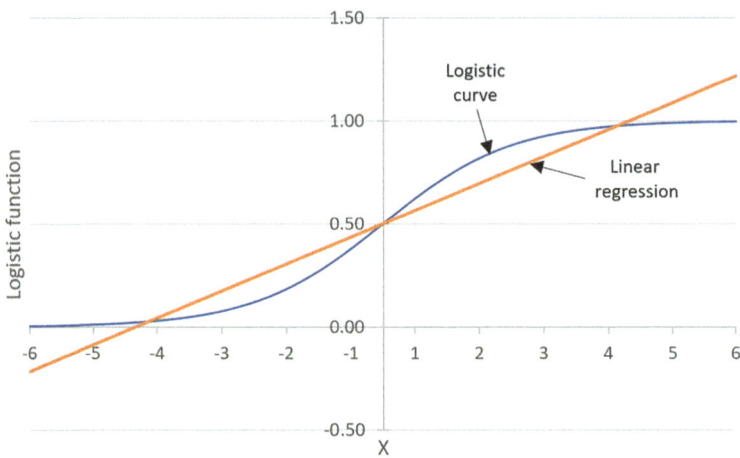

Fig. 7.1   Logistic curve versus linear regression line

This chapter also includes a detailed discussion of the metrics used to assess predictive models where the target variable is binary or multi-level categorical. This discussion is relevant to other prediction models used for categorical targets covered in later chapters.

7.1   Intuition About Binary Logistic Regression

The illustration in Fig. 7.2 presents a conceptual view of how a data set is used in logistic regression.

The y_i values represent the observed values in Fig. 7.2. The x_{ij} predictors can be continuous (e.g., age) or categorical (e.g., sex) variables. Logistic regression predicts the binary outcomes in two steps. First, the probabilities of 1's and 0's (the pr_i's) are estimated for each observation. Second, a threshold probability level is used to convert the probabilities to binary predictions.

One way of looking at the logistic model is to consider the example of consumers who may or may not attend a particular concert. One can imagine the probability of attending the event underlying each consumer decision. Some people will have low probabilities, others moderately high probabilities, and still others very high probabilities. A few with low chances will attend, and most customers with high probabilities will attend. But these probabilities are not observable. The only data that can be collected on these consumers is the binary outcome of whether they attend the concert or do not attend the concert.

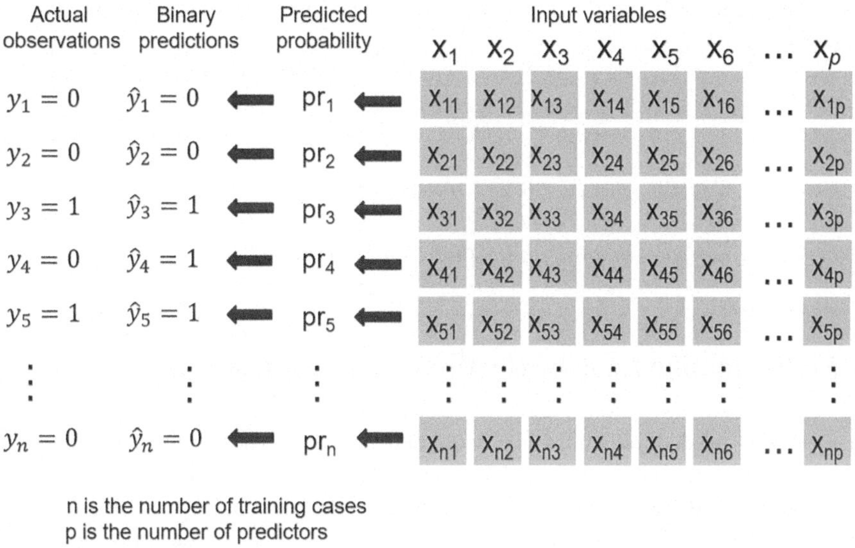

Actual observations	Binary predictions	Predicted probability	Input variables							
			X_1	X_2	X_3	X_4	X_5	X_6	...	X_p
$y_1 = 0$	$\hat{y}_1 = 0$	pr_1	X_{11}	X_{12}	X_{13}	X_{14}	X_{15}	X_{16}	...	X_{1p}
$y_2 = 0$	$\hat{y}_2 = 0$	pr_2	X_{21}	X_{22}	X_{23}	X_{24}	X_{25}	X_{26}	...	X_{2p}
$y_3 = 1$	$\hat{y}_3 = 1$	pr_3	X_{31}	X_{32}	X_{33}	X_{34}	X_{35}	X_{36}	...	X_{3p}
$y_4 = 0$	$\hat{y}_4 = 1$	pr_4	X_{41}	X_{42}	X_{43}	X_{44}	X_{45}	X_{46}	...	X_{4p}
$y_5 = 1$	$\hat{y}_5 = 1$	pr_5	X_{51}	X_{52}	X_{53}	X_{54}	X_{55}	X_{56}	...	X_{5p}
⋮	⋮	⋮	⋮	⋮	⋮	⋮	⋮	⋮		⋮
$y_n = 0$	$\hat{y}_n = 0$	pr_n	X_{n1}	X_{n2}	X_{n3}	X_{n4}	X_{n5}	X_{n6}	...	X_{np}

n is the number of training cases
p is the number of predictors

Fig. 7.2 Typical data structure for binary logistic regression

7.2   Modeling Probabilities

Even though the probabilities cannot be observed, it is helpful to conceptualize a model where the probabilities are a function of the variables describing the customers (7.1).

$$p(y_i = 1) = f\left(\beta_0 + \beta_1 x_{i1} + \cdots + \beta_j x_{ij} + \cdots \beta_p x_{ip} \right) \qquad (7.1)$$

Where y_i is the probability that customer i attends the concert
the x_{ij} are variables measured on each customer, and
the β_j are coefficients on predictors.

The probabilities estimated using Eq. (7.1) can be converted into binary outcomes using a *threshold* value, as shown in Eq. (7.2). The default threshold is usually set at 0.5 but can be varied up or down in situations where the consequences of false positives (incorrectly predicting a binary one) and false negatives (incorrectly predicting a binary zero) differ greatly. Examples of adjusting the threshold will be discussed later.

$$if\ p(y_i = 1) \geq threshold, then\ \widehat{y}_i = 1, else\ \widehat{y}_i = 0 \qquad (7.2)$$

The function $f(\ )$ in Eq. (7.1) is the logistic function, which equals the natural logarithm of odds. Odds for each observation are computed as the natural logarithm of the ratio of the probability that $y_i = 1$ to the probability that $y_i = 0$. This function is called the logit (7.3).

$$logit_i = \ln\left(Odds\,|_{y_i}\right) = \ln\left(\frac{p\,|_{y_i=1}}{p\,|_{y_i=0}}\right)$$

$$= \beta_0 + \beta_1 x_{i1} + \cdots + \beta_j x_{ij} + \cdots \beta_p x_{ip}$$

(7.3)

Where $p\,|_{y_i=1}$ *indicates the probability that* y_i *equals* 1 *given the* x_{ij}.

7.3 Estimating Logistic Regression Parameters

A method is needed to estimate the coefficients in Eq. (7.3). As with linear regression models, "training" the logistic model refers to learning the coefficients that generate the predicted values as close to the observed values as possible for all cases in the dataset. Unfortunately, unlike linear regression, no closed-form algebraic solution is available with logistic regression. Instead, an iterative algorithm obtains successive approximations of the model weights (the β_j) using a maximum likelihood function. The algorithm is efficient, and reasonable estimates are almost always found after several iterations. In practice, since an iterative process is used, the final solution is not guaranteed to be globally optimum but is sufficiently close to optimum in most cases.

7.4 Example Using Simulated Data

This first example shows how KNIME can fit a logistic model with two predictors. Simulated data for this example was created with two continuous predictors, x_1 and x_2. The target variable y_i is binary with two classes, 0 and 1. Fifty cases were created, 23 with $y_i = 1$ and 27 with $y_i = 0$. Figure 7.3 has the KNIME workflow, and Table 7.1 contains the node descriptions for the workflow.

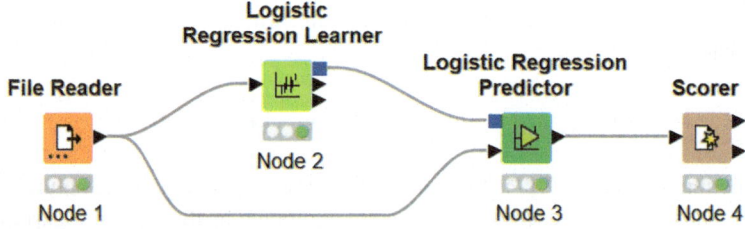

Fig. 7.3 Workflow for simulated data

Table 7.1 Node descriptions for the Fig. 7.3 workflow

Nodes	Descriptions	Input and output ports
Node 1 File reader	Read file: SimulatedDataAnalysisForKNIME.csv	<u>Output port:</u> data table, 60 rows by 3 columns
Node 2 Logistic regression learner	Create the logistic model Settings: Target column: Y Select solver: iteratively reweighted least squares Include: predictors x_1 and x_2	<u>Input port:</u> data table from the output port of node 1 <u>Upper output port:</u> logistic regression model <u>Middle output port:</u> coefficients and statistics from logistic model <u>Lower output port:</u> model and learning properties
Node 3 Logistic regression predictor	Predict the target variable using the logistic model on the simulated data set Settings: Check: custom prediction column name: left predict Check: append columns with predicted probabilities	<u>Upper input port:</u> logistic model from node 2 upper output port <u>Lower input port:</u> data table from output port of node 1 <u>Output port:</u> original test data table plus predicted values
Node 4 Scorer	Produce accuracy statistics and a confusion matrix for the logistic model training set Scorer: First column: Y Second column: Prediction(Y)	<u>Input port:</u> data table from node 3 output port <u>Upper output port:</u> confusion matrix <u>Lower output port:</u> accuracy statistics

Table 7.2 Final estimated logistic regression coefficients

Coefficients	KNIME estimates
$\hat{\beta}_1$	6.496
$\hat{\beta}_2$	−6.870
$\hat{\beta}_3$	−3.375

The Logistic Analysis Results

Equation (7.4) shows the resulting logit model with the coefficients repeated in Table 7.2.

$$logit_i = \ln\left(Odds \mid_{y_i}\right) = \ln\left(\frac{p \mid_{y_i=1}}{p \mid_{y_i=0}}\right) = 6.496 - 6.870 x_{i1} - 3.375 x_{i2} \qquad (7.4)$$

Estimating Probabilities

Probabilities for each observation in the data set are obtained by applying an exponential transformation e^x to both sides of Eq. (7.4), resulting in Eq. (7.5). Solving for $p|_{y_i=1}$ yields the prediction Eq. (7.6).

$$e^{\ln\left(\frac{p|_{y_i=1}}{p|_{y_i=0}}\right)} = e^{6.496-6.870x_{i1}-3.375x_{i2}} \tag{7.5}$$

$$\hat{p}\,|_{\hat{y}_i=1} = \frac{1}{1+e^{-(6.496-6.870x_{i1}-3.375x_{i2})}} \tag{7.6}$$

Since the logistic regression model was used to make predictions of the binary target, the next step is to evaluate how well the model performed. A partial table of the predictions using Eq. (7.6) is in Table 7.3. Table 7.4 shows a crosstabulation of all 60 predictions by the actual data, which indicates accuracy.

Beyond measuring accuracy, analysts are interested in interpreting the model to make sure it makes sense. Some ways to do this are discussed in the next section.

Table 7.3 Partial results from logistic regression on simulated data

RowID	x_{1i}	x_{2i}	y_i	Prob($y_i = 1$)	Prob($y_i = 2$)	Predicted y_i
Row0	−0.45	1.09	1	0.997	0.003	1
Row1	−0.56	1.51	1	0.995	0.005	1
Row2	0.06	1.01	1	0.936	0.064	1
Row3	−0.07	1.45	1	0.889	0.111	1
...	...	...	...	...	...	...
Row45	0.58	5.06	0	0.000	1.000	0
Row46	0.44	5.67	0	0.000	1.000	0
Row47	0.47	2.92	0	0.001	0.999	0
Row48	−0.28	1.27	1	0.984	0.016	1
Row49	0.15	1.88	0	0.293	0.707	0

Table 7.4 Prediction accuracy for the logistic model with simulated data

	Predicted Y values	
Actual Y	0	1
0	26	1
1	2	21
Accuracy = (26 + 21)/(26 + 1 + 2 + 21) = 0.94		

7.5   The Nonlinearity of Logistic Regression Coefficients

One practical question related to interpreting a logistic model is, "How do changes in each predictor affect the value of the target variable?" This question is easily answered in ordinary linear regression (OLS). For example, in an OLS model with a continuous target variable, y, a coefficient on a predictor variable of 0.15 means that for every one-unit increase in that variable, the target variable y, increases by 0.15. Logistic regression is more complex since the model is linear only with log odds (or logit) changes. A predictor in a logistic model with a coefficient of 0.15 results in a 0.15 increase in the logit, but it is not intuitive what this means.

   The following table and charts illustrate the effects on probability with changes in a predictor in logistic regression. The changes in probability as predictors change is not a constant due to the non-linear model. The data set and results are from the example that was discussed earlier. Table 7.5 gives the differences in values of x_1 and x_2 and the changes in probabilities.

Change in X_1 (−0.07 to +0.03) with $X_2 = 1.50$

In Fig. 7.4, the predictor X_2 is held constant at 1.50, and X_1 varies from −0.07 to +0.03, an increment of 0.10. The chart shows that the increase of 0.10 in X_1 results in a change in probability of −0.10. (Recall that the coefficient on variable X_1 in the logistic regression model was negative.)

Change in X_1 (+0.170 to +0.270) with $X_2 = 1.50$

In Fig. 7.5, the predictor X_2 is held constant at 1.50, and X_1 varies from 0.170 to 0.270, an increment of 0.10. The chart shows that the increase of 0.10 in X_1x1 results in a change in probability of −0.10.

Table 7.5 Changes in probabilities as X_1 and X_2 are changed

Figure	X_1	X_2	Change in probabilities
Figure 7.4	Change in X_1 from −0.07 to 0.03 = 0.10	1.5	Change in probability from 0.87 to 0.77 = −0.10
Figure 7.5	Change in X_1 from 0.17 to 0.27 = 0.10	1.5	Change in probability from 0.57 to 0.40 = −0.17
Figure 7.6	Change in X_1 from −0.07 to 0.03 = 0.10	2.0	Change in probability from 0.57 to 0.39 = −0.18
Figure 7.7	Change in X_1 from 0.17 to 0.27 = 0.10	2.0	Change in probability from 0.19 to 0.11 = −0.08

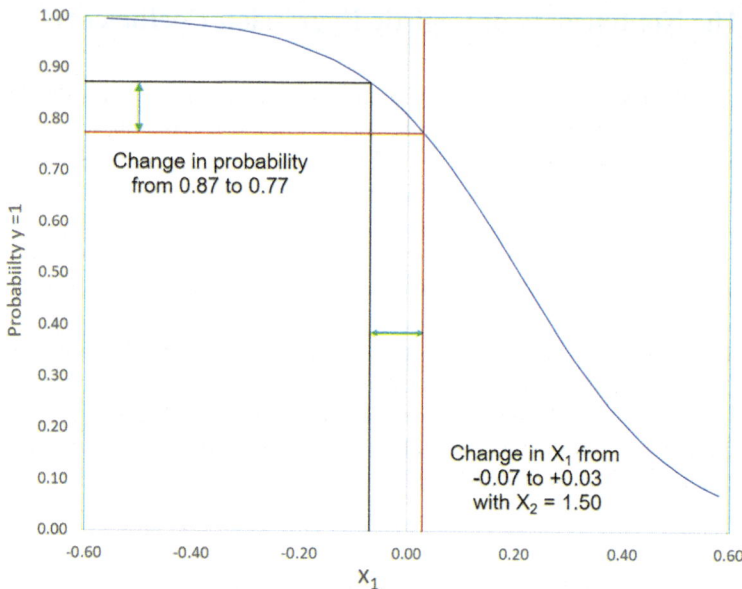

Fig. 7.4 Change in probability as X_1 increases from -0.07 to $+0.03$ with $X_2 = 1.50$

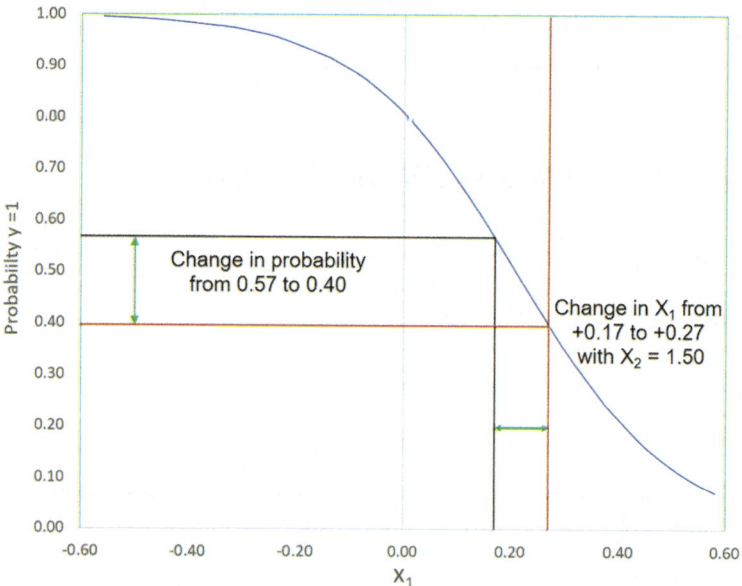

Fig. 7.5 Change in probability as X_1 increases from $+0.17$ to $+0.27$ with $X_2 = 1.50$

Change in X₁ (−0.07 to +0.03) with X₂ = 2.00

Changes in the probability with changes in X_1 also depend on the value of X_2. Figure 7.6 shows the effect of the X_1 change increases from −0.07 to +0.03 with X_2 held constant at 2.00. This results in a change in probability from 0.57 to 0.39.

Change in X₁ (−0.07 to +0.03) with X₂ = 2.00

Changes in the probability with changes in X_1 also depend on the value of X_2. Figure 7.7 shows the effects of a change in X_1 from +0.17 to +0.27 with X_2 held constant at 2.00. The change in probability is −0.08.

The examples shown in the table and the figures demonstrate the difficulty of determining the effect of predictors on the probability that the target variable equals one. The logistic model *is* linear in the log odds, as shown in Eq. (7.3). Each coefficient indicates that a one-unit change in a predictor produces a constant change in the log odds. The changes in log odds also do not depend on the values of the other predictors. But the problem is that changes in log odds have little substantive meaning. The only interpretable statement that can be made about the logistic coefficients is that increases in predictors with positive coefficients result in increases in the log odds (and probabilities) and vice versa.

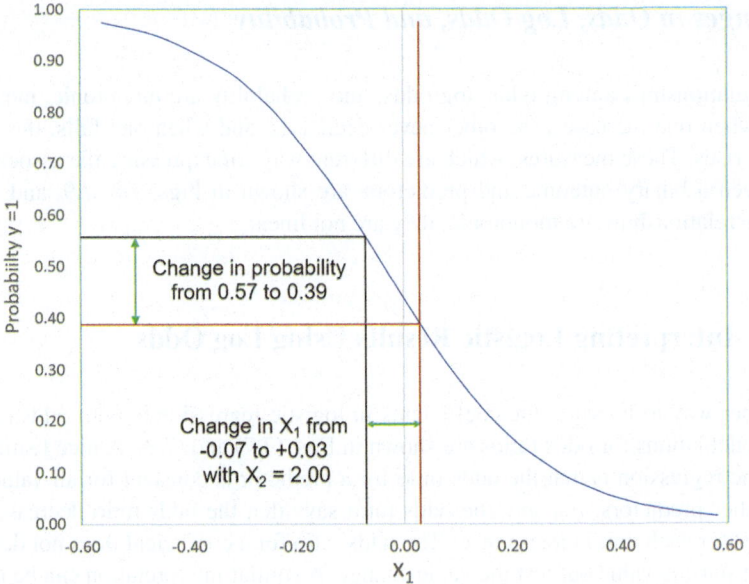

Fig. 7.6 Change in probability as X_1 increases from −0.07 to +0.03 with $X_2 = 2.00$

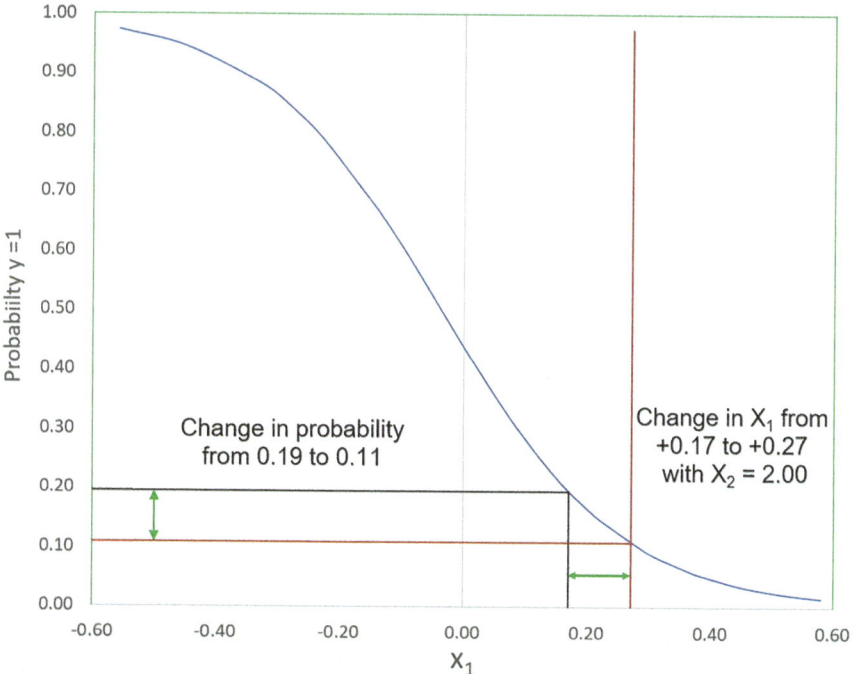

Fig. 7.7 Change in probability as X_1 increases from +0.17 to +0.27 with $X_2 = 2.00$

Changes in Odds, Log Odds, and Probability

The relationships among odds, log odds, and probability are monotonic, meaning that when one increases, the other never decreases, and when one falls, the other never rises. These measures, which are different ways of expressing the association between a binary outcome and predictors, are shown in Figs. 7.8, 7.9, and 7.10. While relationships are monotonic, they are not linear.

7.6 Interpreting Logistic Results Using Log Odds

Another way to interpret the coefficients in logistic regression is with odds ratios. The calculations for odds ratios are shown in Eqs. (7.7) and (7.8). A nice feature of logistic regression is that the odds ratio for a predictor is constant for all values of the other predictors. For $\hat{\beta}_1$, the odds ratio says that the odds ratio decreases by 0.0010 for each unit increase in x_1. The odds ratio for a coefficient does not depend on the starting value but just the value change. A similar interpretation can be made for the odds ratio of $\hat{\beta}_2$. The formulas for the odds ratios of $\hat{\beta}_1$ and $\hat{\beta}_2$ are shown in the equations.

Fig. 7.8 Odds vs. log odds

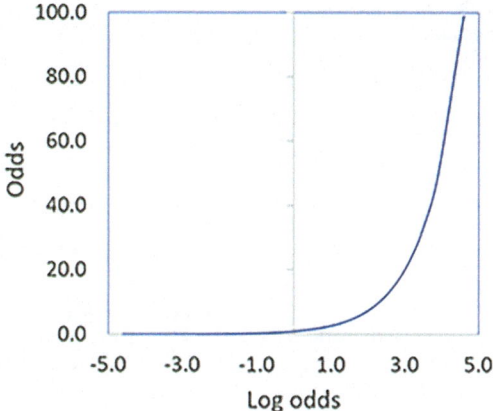

Fig. 7.9 Probability vs. odds

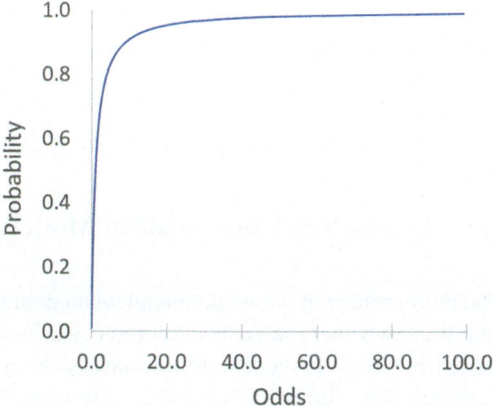

Fig. 7.10 Probability vs. log odds

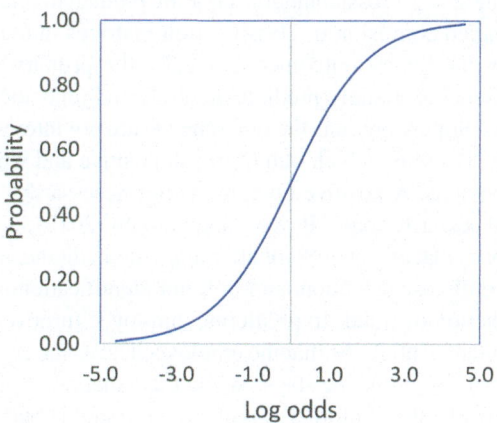

Table 7.6 Calculating
odds ratios

Coefficients	Estimates	Odds ratio
$\hat{\beta}_1$	−6.870	$e^{-6.870} = 0.0010$
$\hat{\beta}_2$	−3.375	$e^{-3.375} = 0.0342$

$$\text{odds ratio for } \hat{\beta}_1 = \frac{\text{odds}\left(y = 1, \text{given } x_1 = \text{any value} + 1\right)}{\text{odds}\left(y = 1, \text{given } x_1 = \text{any value }\right)} = 0.0010 \qquad (7.7)$$

$$\text{odds ratio for } \hat{\beta}_2 = \frac{\text{odds}\left(y = 1, \text{given } x_2 = \text{any value} + 1\right)}{\text{odds}\left(y = 1, \text{given } x_2 = \text{any value }\right)} = 0.0342 \qquad (7.8)$$

The odds ratios are easily calculated by exponentiating the coefficients on the predictors using the base of the natural logarithm, e, as shown in Table 7.6.

Interpreting the coefficients in a logistic regression model is frequently helpful in understanding whether a model makes sense. However, the predictive performance of classification models is nearly always the most important kind of assessment that is made.

7.7 Evaluating Classification Models

The performance of a logistic model when predicting a binary target can ultimately determine whether the goals of a project are met and if the model is useful. As discussed in Chap. 1, models do not have to be perfectly accurate to be useful, but assessing how well predicted values correspond to known classifications is critical.

The basis for several performance metrics with binary classification models is the 2 × 2 crosstabulation table of predicted versus actual values. These tables are called confusion or classification matrices in the prediction analytics context. One of the binary outcomes is usually the primary concern in the application. If the model accurately predicts the preferred outcome, it is a true positive. If the model accurately predicts the outcome we are not interested in, it is labeled a true negative.

Deciding which output to call positive and which to label negative is not always obvious. A positive outcome is not necessarily "good," nor is a negative outcome necessarily "bad." Provost and Fawcett (2013) suggest that positive outcomes are the ones that are "worthy of attention". In medicine, a positive prediction outcome might be disease detection, and in fraud identification, a positive outcome could be a prediction of fraud. In predicting employee turnover, the positive outcome, the one of interest, might be that the employee leaves the company, while a negative outcome is that the employee stays. While this labeling might not be intuitive, it is essential to clearly state before a classification model is built to define the meaning of positive and negative outcomes. Many of the measures described below that are used to assess performance and interpret results in confusion matrices depend on clear definitions of which outcome is considered positive and which is regarded as negative.

Table 7.7 Cells in a confusion matrix

	Prediction	
Actual	Negative	Positive
Negative	True negative	False positive
Positive	False-negative	True positive

Confusion Matrices

A confusion matrix can have four labeled cells, as shown in Table 7.7. The table shows the actual values in rows and the predicted values in the columns. The confusion matrices produced by the Scorer node in KNIME conform to this structure. (Other software may reverse this arrangement, putting the actual in columns and the predicted in rows, so it is important to check how a specific software tool displays confusion matrices.)

Several measures are frequently used to assess model performance using a 2 × 2 table, all provided by the Scores node in KNIME. These are:

- *Accuracy* = (True positive + True negative)/Total cases. (Also known as the "fraction correct.")
- *Sensitivity* = (True positive)/(True positive + False negative). (Also known as "recall.")
- *Specificity* = (True negative)/(True negative + False positive). (Also known as "true negative rate.")
- *Precision* = (True positive)/(True positive + False positive). (Also known as "positive predictive value.")
- *F-score* = 2 × [(sensitivity × precision)]/[(sensitivity + precision)].

Which Metrics to Use with Confusion Matrices?

Since every application is different, there is no single correct classification metric for all situations. Accuracy is valuable when the numbers of false positives and false negatives are approximately equal, and the cost or seriousness of false positives and false negatives are similar. Consider the hypothetical results for predicting the Covid-19 virus and SPAM email in Tables 7.8, 7.9, 7.10 and 7.11, which show hypothetical results for two prediction models. It is assumed that the positive outcome for Covid-19 is correctly detecting the virus, and for SPAM, the positive outcome is correctly identifying an email as SPAM.

The performance metrics for the four tables are shown with formulas in Table 7.12. Accuracy in all four tables = (2000 + 5000)/(2000 + 10 + 400 + 5000) = 94.4%, yet the models are not equally favorable. Accuracy is not a good measure for these two situations since the consequences of false positives and false negatives differ.

For the Covid-19 tests, it is more important to avoid classifying cases of actual Covid than not having Covid. Therefore, model 1 is preferred to model 2 because only 10 Covid cases were mistakenly classified as Not Covid. Model 1 has a greater number of false positives, and despite possibly stressing the patient and causing further tests, this is a less severe error. In general, if false positives are not as serious as false negatives, sensitivity is a good metric, which is 0.998 in this example.

Table 7.8 Covid-19 model 1

	Predicted	
Actual	Not Covid	Covid
Not Covid	2000	400
Covid	10	5000

Table 7.9 Covid-19 model 2

	Predicted	
Actual	Not Covid	Covid
Not Covid	2000	10
Covid	400	5000

Table 7.10 SPAM model 1

	Predicted	
Actual	Not SPAM	SPAM
Not SPAM	2000	400
SPAM	10	5000

Table 7.11 SPAM model 2

	Predicted	
Actual	Not SPAM	SPAM
Not SPAM	2000	10
SPAM	400	5000

Table 7.12 Metrics for Covid-19 and SPAM confusion matrices

Model	Accuracy	Precision	Sensitivity	Specificity	F-Score
Covid-19 model 1	0.945	0.926	0.998	0.833	0.908
Covid-19 model 2	0.945	0.998	0.926	0.995	0.959
SPAM model 1	0.945	0.926	0.998	0.833	0.908
SPAM model 2	0.945	0.998	0.926	0.995	0.959

For the SPAM tests, it is more important that the model does not send non-SPAM messages into the SPAM folder of an email program. If this happens, the email user might never see critical messages. Precision is a good metric if false positives are more serious than false negatives. So, model 2 is better with a precision equal to 0.998.

Another frequently used assessment tool for binary prediction models, the ROC curve, is discussed in the next section. The ROC curve is useful for comparing predictive models visually.

ROC Curves

ROC (receiver operating characteristic) curves show the tradeoff that must occur between true positives and false positives in a predictive model. There is always such a tradeoff in a single predictive model. The true positive rate is sensitivity, and the false positive rate is obtained by taking one minus specificity.

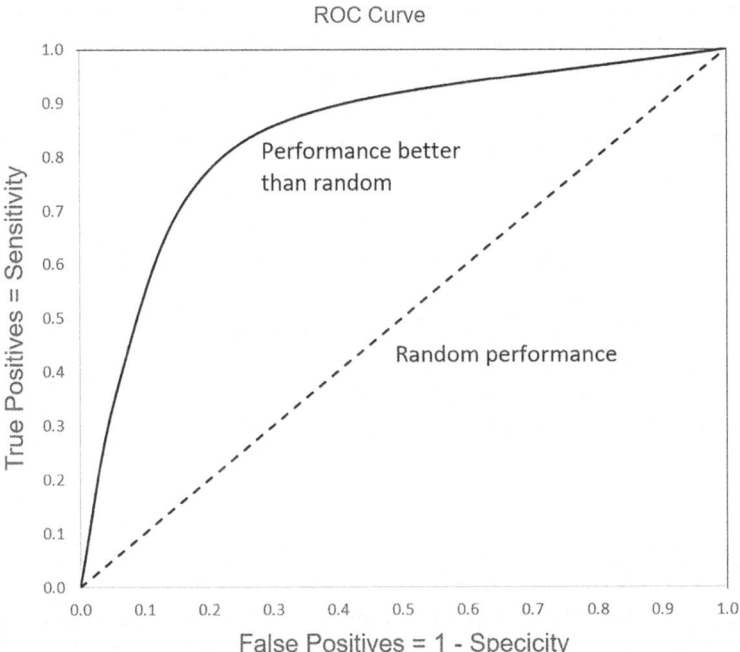

Fig. 7.11 Example of an ROC curve

Figure 7.11 shows a typical ROC curve. If observations were randomly assigned as positive or negative, the performance would be shown as the dashed line in the ROC diagram. (If performance is below the dashed line, the performance would be worse than random.) The gap between the dashed line and the solid ROC curve is of interest in ROC charts. The larger the gap and the closer the ROC solid line is to the upper left corner, the better the model.

The area under the ROC curve is reported as a summary measure of model performance. The area has a maximum of 1.0 since both the False Positive and True Positive axes range from 0.0 to 1.0. If the area is close to 0.50, then the model does not do much better than randomly assigning observations to positive or negative outcomes.

ROC curve history

ROC curves were developed during World War II. These were used to guide radar operators to discern between enemy and Allied aircraft on the radar screens. If the discernment was too sensitive, then while many enemy planes would be correctly identified, Allied aircraft would also be mistakenly identified as enemies. If the opposite occurred, the radar operator would make a few mistakes, incorrectly identifying Allied aircraft as enemies but also missing many enemy aircraft (Kothawade, 2021).

ROC curves are derived by changing the threshold probability of assigning observations to positive or negative outcomes. The default threshold level is usually set to 0.50, but it can be varied between 0.0 and 1.0. Each point on the ROC curve corresponds to one decision threshold value. The left panel in Fig. 7.12 superimposes several decision threshold values on the ROC chart. The ROC chart traces the threshold range from 0.0 to 1.0, clearly showing the tradeoff between true and false positives.

The tradeoff idea is also demonstrated in the right panel of Fig. 7.12. Assume that the goal is to have 90% sensitivity. This can be achieved by setting the threshold very low. To determine the threshold from the chart, a line can be drawn on the y-axis at 0.90 until it intersects the ROC curve. The intersection occurs at a threshold of about 0.10. A line can then be dropped from the ROC curve from the 0.10 point down to the false positive rate on the x-axis. The dropped line intersects the x-axis midway between 0.40 and 0.50. This means that while sensitivity is relatively high at 90%, false positives will be more than 40%. If this rate of false positives is unacceptable, it can be reduced by increasing the threshold, which will reduce the sensitivity or True Positives. The only way to avoid this tradeoff is to develop a more accurate model with an ROC curve above that shown in Fig. 7.12.

ROC curves also provide a convenient way of comparing the performance of two or more predictive models. Figure 7.13 compares the ROC curves of the two models. The model represented by the solid black line dominates the performance of the model represented by the dashed line. It sometimes happens that the curves from two or more ROC curves cross, as shown in Fig. 7.14. In this case, the model represented by the solid line is better when the True Positives are low, and the False Positives are low. At higher levels of True Positives, the model in Fig. 7.14, given by the dashed line, is better.

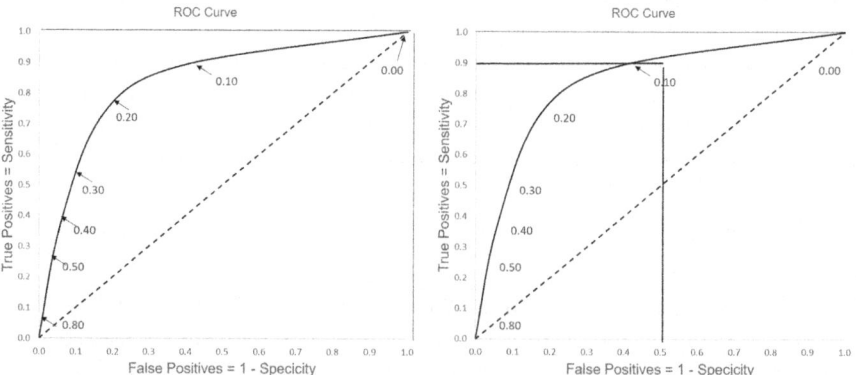

Fig. 7.12 Threshold values on an ROC curve

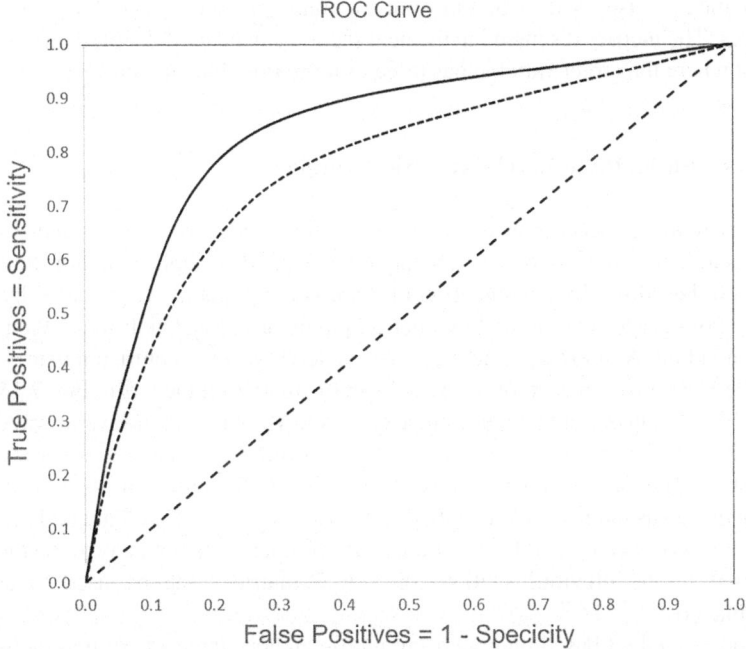

Fig. 7.13 Comparing models using ROC curves

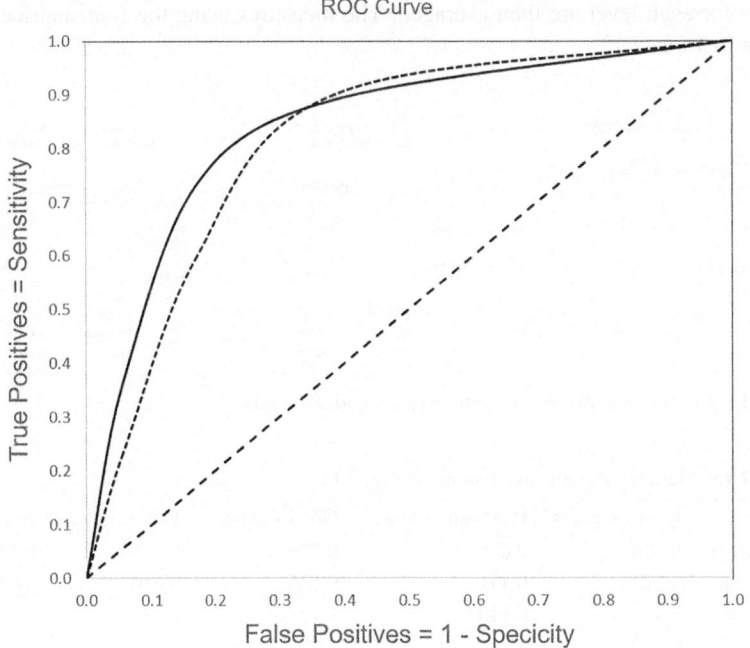

Fig. 7.14 Models with overlapping performance

The metrics discussed so far apply only to binary predictions and 2 × 2 confusion matrices. The metrics discussed in the next section can be used if a predictive model is used where the target variable has three or more possible outcomes.

Metrics with Multiple-Level Categorical Targets

One way to assess performance in cases where the target variable has three or more levels is to reduce the outcome to a binary case. A good example to demonstrate this approach, based on the classification of diamonds by quality, is provided by Bex (2021). Four grades of diamond in order of quality are considered: Ideal, Premium, Good, and Fair. Assume a predictive model yielded the 4 × 4 confusion matrix with hypothetical entries in each of the 16 cells in the matrix on the left in Fig. 7.15.

The 4 × 4 matrix can be reduced to a 2 × 2 confusion matrix in several ways. For example, collapsing the 4 × 4 matrix can be achieved if one outcome is particularly interesting. The matrix shown in the center of Fig. 7.15 focuses on predicting Ideal diamonds versus the rest. In the matrix on the right in Fig. 7.15, Ideal and Premium diamonds versus Good and Fair diamonds are compared. In both cases, the metrics discussed in the previous section for 2 × 2 matrices can be used to assess performance.

Another method based on collapsing multi-category targets is to find the sensitivity, precision, and F1 score by forming 2 × 2 matrices using each level versus the others. The example above requires that 2 × 2 matrices be obtained for Ideal versus others, Premium versus others, Good versus others, and Fair versus others. The results for each level are then averaged. The measures using the four matrices are shown in Table 7.13.

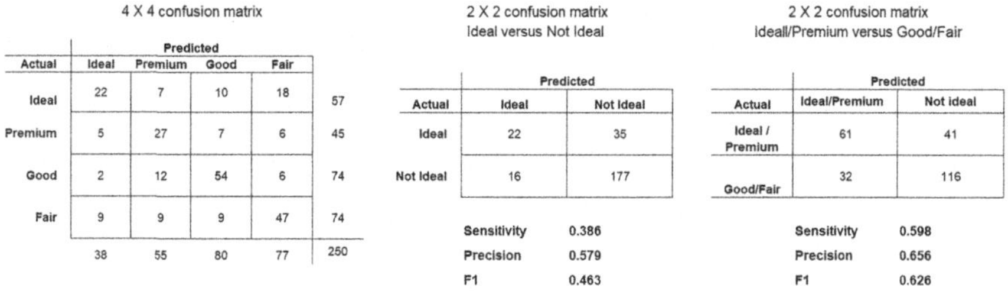

Fig. 7.15 Confusion matrices for multi-level categorical targets

Table 7.13 Metrics for confusion matrices in Fig. 7.15

Metric	Ideal vs others	Premium vs others	Good vs others	Fair vs others	Average
Sensitivity	0.386	0.600	0.730	0.635	0.588
Precision	0.579	0.491	0.675	0.610	0.589
F1 score	0.463	0.540	0.701	0.622	0.582

If collapsing to binary tables is not reasonable, overall accuracy might be used. Overall accuracy could be calculated by summing the entries on the diagonal of the 4×4 matrix in Fig. 7.15 and dividing them by the total number of observations in the table. However, computing accuracy in this way is not desirable because the metric will not consider the distribution of misclassifications. Logically, the classifications (and misclassifications) of levels with a large number of observations should be weighted more heavily in evaluating performance.

One metric that does consider the distribution of multi-level classification is Cohen's Kappa (Pykes, 2020), which is provided in the classification metrics produced with KNIME's Score node. The formula for Cohen's Kappa compares the observed accuracy with the prediction model with the expected accuracy under random chance (Eq. 7.9).

$$\text{Cohen's Kappa} = \frac{\left(observed\ diagonal\ obs - expected\ diagonal\ obs\right)}{\left(Total\ observations - expected\ diagonal\ obs\right)} \tag{7.9}$$

Calculating Cohen's Kappa requires estimating the expected values along the diagonal of the confusion matrix, which is explained in the Appendix to this chapter. Kappa only considers accuracy without assessing sensitivity, precision, and other metrics.

7.8 Example: Predicting Employee Retention Using Logistic Regression

An illustration of logistic regression to predict employee retention (and its inverse, turnover) is developed here. According to the 2020 retention report from the Work Institute,[1] more than 27% of U.S. employees voluntarily left their jobs in 2020, resulting in an estimated cost of $630 billion. Recruitment and training of new employees plus loss of productivity contribute to the costs. The study furthermore reported that more than three-fourths of the turnover was preventable.

Predictive analytics can enable employers to identify employees at risk of leaving. For those at risk that the organization wants to retain, the employer can take preemptive steps to reduce the chances of turnover. To build a predictive model, a data set with each entry containing employee characteristics and whether the employee left is needed. A data set on employee turnover with 14,999 rows and nine columns is available from Kaggle.[2] Variable descriptions are in Table 7.14.

Before running the logistic regression, an exploratory analysis of the data set was conducted. The analysis examined each continuous variable for outliers or extreme skewness and each categorical or string variable for sparse frequencies. The

---

[1] https://info.workinstitute.com/en/retention-report-2020

[2] HR Analytics. https://www.kaggle.com/code/jacksonchou/hr-analytics

Table 7.14 Variables in the HR turnover data set

Variable	Description
Satisfaction	Level scored 0 to 1
Evaluation	Last evaluation rating scored 0 to 1
Projects	Number of projects completed while at work
Hours	Average monthly hours at the workplace
Years	Number of years spent in the company
Promotion	Whether the employee was promoted in the last 5 years (yes, no)
Department	Sales, technical, support, IT, product_mgt, marketing, ResAndDevel, accounting, hr, management
Salary	Relative level of salary: low, medium, high
Left	Whether the employee left the workplace or not (left, did not leave); target variable

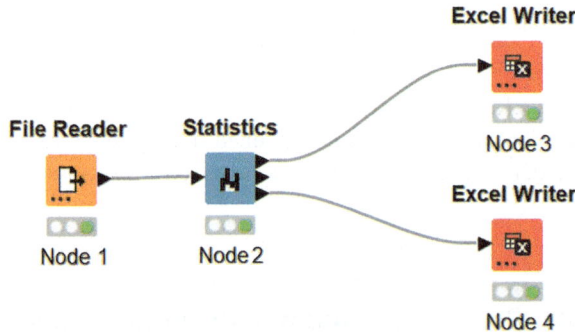

Fig. 7.16 Workflow for exploration of employee turnover data

workflow for the exploratory analysis is in Fig. 7.16, and the node descriptions are in Table 7.15.

In Table 7.16, all the continuous variables are shown to have low skewness except YearsAtCompany, which is moderately skewed. There are no missing values for these variables. Table 7.17 has the frequencies of values for the categorical variables.

Since YearsAtCompany was moderately skewed, it can adversely affect the performance of a classification model such as logistic regression. So, a log-transformed was applied to the variable. The skewness of YearsAtCompany was reduced from 1.853 to 0.589 after the transformation.

Another issue with the data, apparent in Table 7.17, is the imbalance in the target variable, Left vs. Did not leave. There are more than three times the number of observations where the individual did not leave as did leave. This imbalance tends to affect the classification by disproportionately assigning the outcome predictions to the majority level. With extreme levels of imbalance, the model can sometimes assign all cases to the majority level, resulting in a useless predictive model. One approach to remedying this is to oversample the minority level to balance the data.

Table 7.15 Node descriptions for the Fig. 7.16 workflow

Nodes	Descriptions	Input and output ports
Node 1 File reader	Read Employee_turnover.csv.	Output port: data table with employee turnover with 14,999 rows and nine columns
Node 2 Statistics	Calculate statistics such as minimum, maximum, mean, standard deviation, variance, median, overall sum, number of missing values, row count across all numeric columns, and frequency counts for all continuous and numeric variables values. Options: Include all variables Check: calculate median values	Input port: data table from the output port of node 1 Upper output port: statistics on continuous variables Middle output port: histograms for categorical variables (not used in this example) Lower output port: occurrences (frequencies) for categorical variables
Node 3 Excel writer	Write the statistics on continuous variables to an excel file Save excel file: HRStatistics.xlsx	Input port: data table from the upper output port of node 2
Node 4 Excel writer	Write the frequencies of all variables to an excel file Save excel file: HRFrequencies.xlsx	Input port: data table from the lower output port of node 2

Table 7.16 Descriptive statistics on continuous variables in HR data

Variable	Min	Max	Mean	Median	Std. deviation	Skewness	No. missing
Satisfaction	0.09	1	0.613	0.64	0.249	−0.476	0
LastEvaluation	0.36	1	0.716	0.72	0.171	−0.027	0
Projects	2.00	7	3.803	4.00	1.232	0.338	0
AverageHours	96.00	310	201.050	200.00	49.943	0.053	0
YearsAtCompany	2.00	10.0	3.498	3.00	1.460	1.853	0

The SMOTE node can balance the training data to build the model. However, the evaluations using the Scorer nodes use the original unbalanced data for both the training and test sets to make fair assessments of the model's ability to predict new data.

Figure 7.17 has the workflow for the logistic regression on the employee turnover data, and Table 7.18 has the node descriptions.

Results for the Analysis of Employee Turnover

The classification matrices for the training and test data are shown in Tables 7.19 and 7.20. The metrics for the training and test data sets shown in Table 7.21 show that the model's performance for the training and test data subsets is very close.

The analysis was repeated (not shown here) without taking the natural log of YearsAtCompany. An ROC chart was then created to compare the model's performance with the log and without the log transformation. The resulting chart is in Fig. 7.18. The upper curve in the figure is for the analysis with the log transformation (area under the curve = 0.833). This model performed slightly better than the model without the log transformation (area under the curve = 0.820).

Table 7.17 Frequencies of categorical variables in employee turnover data

Variable	Levels	Counts	Relative frequency
Promotion	No	14,680	0.979
	Yes	319	0.021
Department	Sales	4140	0.276
	Technical	2720	0.181
	Support	2229	0.149
	IT	1227	0.082
	Product mgt	902	0.060
	Marketing	858	0.057
	ResAndDevel	787	0.052
	Accounting	767	0.051
	Hr	739	0.049
	Management	630	0.042
Salary	Low	7316	0.488
	Medium	6446	0.430
	High	1237	0.082
Left	Did not leave	11,428	0.762
	Left	3571	0.238

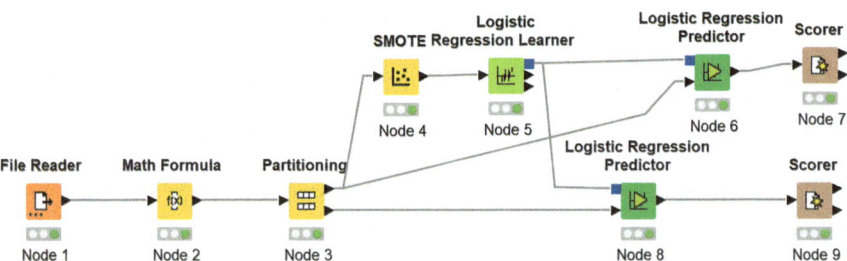

Fig. 7.17 KNIME workflow for logistic regression on employee turnover data

Table 7.18 Node descriptions for the Fig. 7.18 workflow

Nodes	Description	Input and output ports
Node 1 File reader	Read Employee_turnover.csv	<u>Output port:</u> data table with employee turnover with 14,999 rows and 9 columns
Node 2 Math formula	This node can evaluate a mathematical expression based on the row values of a variable using a large assortment of available functions Math expression: ln($YearsAtCompany$) Check replace column: YearsAtCompany	<u>Input port:</u> data table from the output port of node 1 <u>Output port:</u> data table with ln(YearsAtCompany) and 14,999 rows and 10 columns
Node 3 Partitioning	First partition: Select relative: 70% Select stratified sampling: left Check: use random seed: 123	<u>Input port:</u> data table from the output port of node 2 <u>Upper output port:</u> training data with 10,999 rows and 10 columns <u>Lower output port:</u> test data, with 4,500 rows and 10 columns
Node 4 Smote	SMOTE oversamples the input data Settings: Class column: left Check: oversample minority classes Check: enable static seed: 123	<u>Input port:</u> data table from the upper output port of node 3 <u>Output port:</u> oversampled training data with a balanced number of left and did not leave target values, resulting in 15,998 rows and 10 columns
Node 5 Logistic regression learner	Create the logistic model Settings: Target column: left Select solver: iteratively reweighted least squares Include: all predictor variables except replace Years at Company with ln(Years at Company)	<u>Input port:</u> data table from the output port of node 4 <u>Upper output port:</u> logistic regression model <u>Middle output port:</u> coefficients and statistics <u>Lower output port:</u> model and learning properties
Node 6 Logistic regression predictor	Predict the target variable using the logistic model on the training data Settings: Check: custom prediction column name: left predict Check: append columns with predicted probabilities	<u>Upper input port:</u> model from upper output port of node 5 <u>Lower input port:</u> data table from the upper output port of node 3 <u>Output port:</u> Original training data table plus predicted values, with 10,499 rows and 13 columns
Node 7 Scorer	Produce accuracy statistics and a confusion matrix for the logistic model training set Scorer: First column: left Second column: left predict	<u>Input port:</u> data from the output port of node 6 <u>Upper output port:</u> confusion matrix with training data <u>Lower output port:</u> accuracy statistics with training data
Node 8 Logistic regression predictor	Predict the target variable using the logistic model on the test data Settings: Check: Custom prediction column name: left PredictCheck: Append columns with predicted probabilities	<u>Upper input port:</u> model from upper output port of node 5 <u>Lower input port:</u> data table from the lower output port of node 3 <u>Output port:</u> original test data table plus predicted values, 4,500 rows and 13 columns

(continued)

Table 7.18 (continued)

Nodes	Description	Input and output ports
Node 9 Scorer	Produce accuracy statistics and a confusion matrix for the logistic model test set Scorer: First column: left Second column: left predict	Input port: data table from the output port of node 8 Upper output port: confusion matrix with test data Lower output port: accuracy statistics with test data

Table 7.19 Confusion matrix for training data

	Predicted		
Actual	Left company	Did not leave	Total
Left company	2133	367	2500
Did not leave	2034	5965	7999
Total	4167	6332	10,499

Table 7.20 Confusion matrix for test data

	Predicted		
Actual	Left company	Did not leave	Total
Left company	918	153	1071
Did not leave	862	2567	3429
Total	1780	2720	4500

Table 7.21 Performance metrics for training and test data

Metric	Training data	Test data
Accuracy	0.771	0.774
Cohen's Kappa	0.487	0.493
Precision	0.512	0.516
Sensitivity	0.853	0.857
Specificity	0.746	0.749
F-measure	0.640	0.611

7.9 Predictor Interpretation and Importance

Two questions about the employee turnover model from the previous section are: "How can the changes in the predictor variables be interpreted? What is the relative importance of each predictor of employee turnover?" Answering these might provide insight into what steps might be taken to reduce turnover.

The question of predictor (or feature) importance has been studied extensively with multiple regression models. However, comparatively little has been published about measuring predictive importance in logistic regression (Azen & Traxel, 2009). An approach based on "dominance analysis" has been developed and is available in

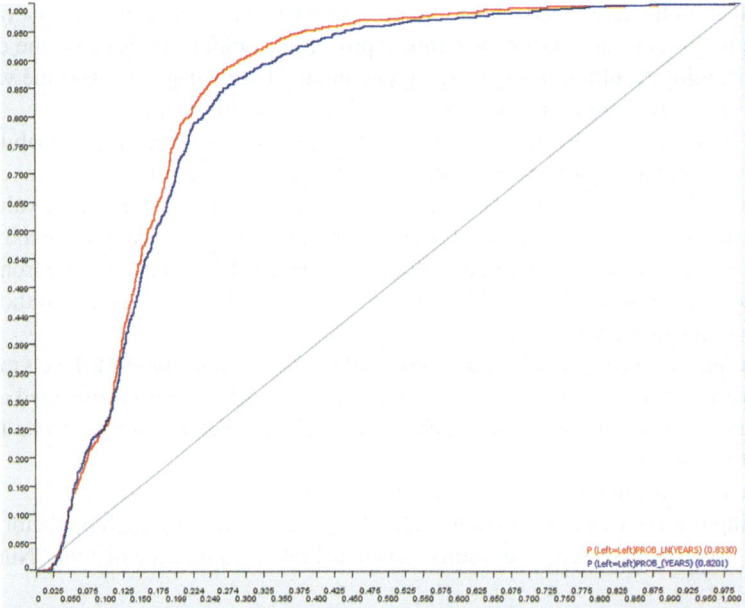

Fig. 7.18 ROC curves to show the effect of a log transformation

Python and R. Since Dominance analysis is not directly available in KNIME, a simpler approach will be discussed in this section.

The coefficients on the predictors cannot be used to infer importance in logistic regression because these coefficients are for the log of the odds. The relative sizes of the coefficients in a logistic model do not indicate predictor importance, even if the predictor variables are normalized. Likewise, the p-values for significance tests on each predictor cannot be interpreted as a measure of importance in a practical sense. A small p-value indicates that the variable has a low variance compared with its magnitude, but it could still have a minor effect on the target variable's probability.

What is usually desired is the impact on the probability of the outcome as changes are made in each predictor. As discussed earlier in the chapter, this is complicated because the effect on the probability of the outcome is not a linear function of changes in a predictor. Furthermore, the probabilities associated with a predictor depend upon the values of the other predictors. Thus, the "importance" of a particular predictor varies based on the range of the predictor being considered and the settings of every other predictor.

An Approximate Method for Predictor Interpretation

In general, for binary logistic models, no approach to interpretation can fully describe the relationship between changes in a predictor value and the probability of the target variable (Long & Frees, 2006). One approximate approach is to examine the effect on the likelihood of the outcome, as each predictor varies over its range. While this seems straightforward, this approach is complicated because the effect on probability resulting from changing one input variable depends upon the values of the other predictors, as was demonstrated earlier in this chapter.

A "baseline" set of values provided a starting point for exploring probabilities. The baseline values for all the continuous predictors were set to their respective means, and the baselines for the nominal variables were set to their modal values.

Next, a range was established for each continuous variable using the maximum and minimum values. Since Department was categorical, the range was set from HR to Management since those in Management were least likely to leave, and those in HR were most likely to leave.

The analysis proceeded by maintaining all predictors at the baseline levels except one that varied over its range. The "importance" of a predictor was computed as the absolute value of the probability difference as the predictor varied from its minimum to its maximum.

The baseline, maximum, and minimum settings are shown in Table 7.22.

Using the values shown in Table 7.22, the effects of varying each predictor over its range can be assessed. The results shown in Table 7.23 indicate that the "Number

Table 7.22 Baseline, maximum, and minimum settings

Variable	Baseline	Max	Min
Satisfaction	0.61	1.00	0.09
Evaluation	0.72	1.00	0.36
Projects	3.80	7.00	2.0
Hours	201.05	310.00	96
Years	3.50	10.00	2.00
Promotion	No	Yes	No
Department	Sales	Management	HR
Salary	Low	High	Low

Table 7.23 Range of probabilities of leaving by predictor

Predictor	Range of probabilities
Number of years spent in the company	0.78
Satisfaction	0.75
Number of projects completed while at work	0.53
Relative level of salary: low to high	0.37
Whether the employee was promoted in last 5 years	0.39
Average monthly hours at the workplace	0.31
Department in which the employee worked	0.22
Last evaluation rating scored	0.15

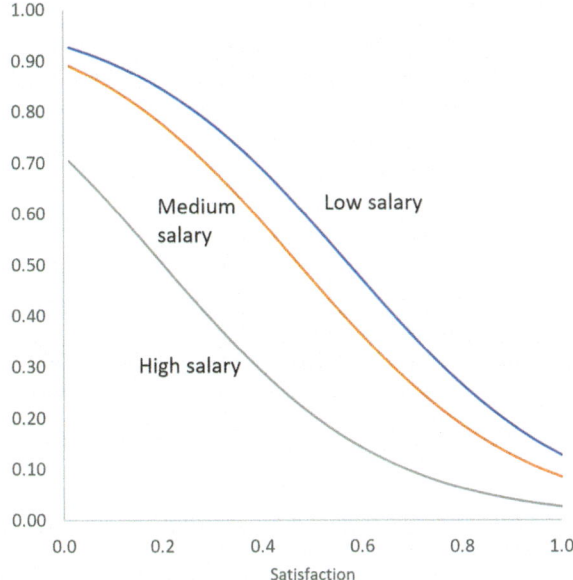

Fig. 7.19 Probability vs. Satisfaction at different Salary levels

of years spent in the company" was most important, followed by "Satisfaction." Note that this method does not fully address the questions of interpretation or importance since only one predictor is varied at a time.

Another way of interpreting predictors is to plot changes in probability by a predictor at various levels of another predictor. Figure 7.19 shows how the probability of leaving the company varies with the level of satisfaction at three salary levels. As might be expected, employees with Low salaries were generally more likely to leave at any level of satisfaction.

This type of analysis is useful when examining the impact of two variables but still does not fully address the question of variable importance.

7.10   Example: Predicting Heart Disease Using Logistic Regression

Logistic regression has been used in many different medical and healthcare studies. This example was first used in Chap. 3 to illustrate KNIME. Now a more detailed examination of the analysis process is shown. The data for this example is from extensive research on heart conditions conducted at the Cleveland Clinic. The goal is to predict the presence of heart disease measured as a binary outcome. The data set here is an abbreviated version of the original study data and contains 270 observations on 14 variables, as shown in Table 7.24.

Table 7.24 Variables in the heart disease data set

Variable	Description
Age	Age in years
Sex	Male; female
Chestpain	Type (4 values as follows)
	Typical angina
	Atypical angina
	Non-anginal pain
	Asymptomatic
Resting_BP	Resting blood pressure (in mm Hg)
Cholesterol	Serum cholesterol in mg/dL
Fasting_sugar_gt_120mg	Fasting blood sugar >120 mg/dl (TRUE, FALSE)
Resting_EKG	Resting electrocardiographic results (three values as follows)
	Normal
	STT_T abnormality
	Left ventricular hypertrophy
Max_heart_rate_achieved	Numeric value of beats per minute
Exercise-induced angina	Yes, no
ST_depression	Depression of trace segment induced by exercise relative to rest
Peak_exercise_ST	The slope of the peak exercise stress test segment (three values as follows)
	Upsloping
	Flat
	Downsloping
Num_vessels_by_ fluoroscopy	Number of major vessels (0–3) colored by fluoroscopy
Thalassemia	Shows how well blood flows into your heart
	Normal
	Fixed defect
	Reversible defect
Heart_disease	Not present; present (target variable)

The KNIME workflow for analyzing the heart disease data is shown in Fig. 7.20, and the node descriptions are in Table 7.25. The nodes for the analysis use the H2O integration with KNIME. The H2O model was used in this example because it includes the ability to perform regularized logistic regression, which will be discussed later in the chapter. H2O is a machine learning platform that provides machine learning tools which are unavailable directly in KNIME. This integration enables KNIME to exploit the latest and most powerful algorithms.

Since several predictor variables were categorical, the KNIME node One to Many created dummy indicator variables for those predictors. The Column Filter was used to drop one dummy indicator for each categorical variable since each k-level categorical value only has k-1 independent levels. This resulted in 21 coefficients to be estimated in the logistic regression.

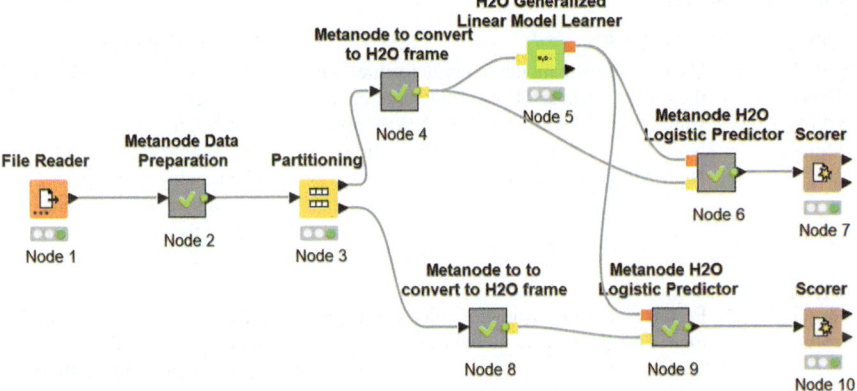

Fig. 7.20 KNIME workflow for logistic regression on heart disease data

Table 7.25 Node descriptions for the Fig. 7.20 workflow

Nodes	Descriptions	Input and output ports
Node 1 File reader	Read file: ClevelandHeartData.csv	Output port: data table with 297 rows and 14 columns
Node 2 Metanode data preparation	This Metanode transforms all categorical variables (except the target variable) into indicator variables coded 0 or 1. For each categorical variable, one indicator is dropped since it is redundant using the column filter Node M1_1 one to many Node M1_2 column filter	Input port: data table from the output port of node 1 Output port: heart data with 297 rows with 5 continuous variables, 15 indicator (0,1) variables, and one categorical target (heart disease present or not present)
Node 3 Partitioning	Set first partition: Select relative: 80% Select stratified sampling: Heart_Disease Check: use random seed: 123	Input port: data table from the output port of node 2 Upper output port: training data (237×21) Lower output port: Test data (60×21)
Node 4 Metanode to convert to H2Oframe	Create local context for running H2O in KNIME Convert KNIME training data table to H2O frame Node M4_1 H2 local context Node M4_2 table to H2O	Input port: data table from the upper output port of node 3 Output port: H2O frame with training data (237×21)
Node 5 H2O generalized linear model learner	Learns a generalized linear model (GLM) classification model using H2O General settings: Target column: Heart_disease Include all other variables Check: use static random seed 123 Algorithm settings: Check set alpha: 0	Input port: data table from the output port of node 4 Upper output port: H2O logistic model Lower output port: model coefficients

(continued)

Table 7.25 (continued)

Nodes	Descriptions	Input and output ports
Node 6 Metanode H2O logistic predictor	Uses the logistic model to predict values with the training data set. Convert the predicted values in the H2O frame to a KNIME table Node M6_1 H2O predictor(classification) Node M6_2 H2O to table	Upper input port: model from the upper output port of node 5 Lower input port: H2O frame with training data (237×21) Output port: KNIME data table with training data and predicted values of heart disease
Node 7 Scorer	Produces accuracy statistics and a confusion matrix for the logistic model training set Scorer First column: Heart_disease Second column: prediction (heart disease)	Input port: data table from the output port of node 6 Upper output port: confusion matrix with training data Lower output port: accuracy statistics with training data
Node 8 Metanode to convert to H2O frame	Create local context for running H2O in KNIME Convert KNIME test data table to H2O frame Node M8_1 H2 local context Node M8_2 table to H2O	Input port: data table from the lower output port of node 3 Output port: H2O frame with training data (60×21)
Node 9 Metanode H2O logistic predictor	Uses the logistic model to predict values with the test data set Convert the predicted values in the H2O frame to a KNIME table Node M9_1 H2O predictor(classification) Node M9_2 H2O to table	Upper input port: model from the upper output port of node 5 Lower input port: data table from the output port of node 8 Output port: Data table with test data and predicted values of heart disease
Node 10 Scorer	Produces accuracy statistics and a confusion matrix for the logistic model test set Scorer: First column: Heart_disease Second column: prediction (heart disease).	Input port: Data table from the output port of node 8 Upper output port: confusion matrix with test data. Lower output port: accuracy statistics with test data

Results for Heart Disease Prediction

The accuracy of predicting heart disease was calculated using the Scorer nodes, and the results are shown in Table 7.26 for the training data and Table 7.27 for the test data. While the training data accuracy is relatively high, the performance on the test data was not as good. This suggests that the model was overfitting the data. This can occur with any predictive model and is especially likely in cases where the number of predictors is large compared to the number of observations. Twenty-one parameters were estimated, and 207 observations were included in this example, so, it is not surprising that the model was overfitting. Regularization can be applied to logistic regression models to mitigate overfitting.

Table 7.26 Prediction results for training data

| | Predicted heart disease | |
Actual	not present	present
Not present	119	9
Present	13	96

Accuracy = (119 + 96)/(119 + 9 + 13 + 96) = .907

Table 7.27 Prediction results for test data

| | Predicted heart disease | |
Actual	not present	present
Not present	26	6
Present	6	33

Accuracy = (26 + 33)/(26 + 6 + 6 + 33) = .830

7.11 Regularized Logistic Regression

Logistic regression can use regularization to improve generalizability, the model's ability to predict new data more accurately. Regularization reduces the chances of overfitting the model to a training set by reducing the chances that the model picks up noise or very unusual data patterns that are unlikely to be found in new data.

Regularization works by changing the learning goal from a single-minded reduction of the prediction error to a dual one where overfitting to the training data is controlled by adding a penalty term. Two penalty terms lead to two types of regularizations: L1 and L2.

L1 regularization (LASSO) adds a penalty term, $\lambda_{L1} \sum_{j=1}^{m} |\hat{\beta}_j|$, to the function to be minimized, as shown in Eq. (7.10). This forces the optimization procedure to assign smaller values to the coefficients in the logistic model. The λ_{L1} parameter controls how much emphasis is given to the penalty term. As λ_{L1} increases, the coefficients will be reduced, forcing some to be zero. A model leaving the most important predictors is created.

$$Funtion\ to\ be\ minimized + penalty = \sum_{i=1}^{n} \log loss_i + \lambda_{L1} \sum_{j=1}^{m} \left|\hat{\beta}_j\right| \qquad (7.10)$$

L2 regularization (ridge regression) adds a slightly different penalty term, $\lambda_{L2} \sum_{j=1}^{m} \hat{\beta}_j^2$. This forces the optimization procedure to assign smaller values to the coefficients in the logistic model, but the coefficients are not necessarily pushed to zero.

By reducing the magnitude of the coefficients, no single predictor is allowed to have a dominant impact on the target value. Even if there are unusual cases in the training data with very large or small values on some predictors, the predicted target values will not change dramatically. Hence, the model becomes more stable and generalizable to new data. The λ_{L1} and λ_{L2} parameters are positive numbers set by the analyst training the model, and they work as levers to adjust both L1 and L2 regularization levels.

Applying Regularization to the Heart Disease Data

Regularization was applied to the heart disease model using the same workflow as in Fig. 7.20 and Table 7.25. The only changes were in the algorithm settings. Two parameters must be set to obtain regularization, alpha and lambda in the H2O module. Table 7.28 shows how to produce L1 and L2 regularization. (The alpha setting in H2O is a "mixing" parameter that can balance L1 and L2 regularization. When the two are balanced, this is known as Elastic Net regularization. This was not used in the present example.)

Both L1 and L2 regularization were applied and compared with the model without regularization. The results are summarized in Table 7.29.

Both the L1 and L2 models reduced the amount of overfitting compared with the "no regularization" model. The L1 regularization simplified the model considerably by setting all but four coefficients on the 21 parameters to zero (Table 7.30). However, this also resulted in lower training data accuracy. The L2 model achieved higher accuracy than the L1 regularization with the test data. The L2 model reduced the sizes of the coefficients but did not set any to zero. The L2 regularization also showed a somewhat greater difference between the training and test data accuracies. The coefficients with no regularization, also shown in Table 7.30, are larger than those from the L1 and L2 regularization.

Table 7.28 Settings in H2O for logistic regression

Type of regularization	Alpha setting	Lambda setting
L1 LASSO	1	>0
L2 ridge regression	0	>0
No regularization	Ignored	0

Table 7.29 Accuracy of predicting heart disease for the training and test data

	No regularization	L1 regularization	L2 regularization
Accuracy – training	0.907	0.831	0.873
Accuracy – test	0.800	0.800	0.817
Overfitting	0.907–0.800 = 0.107	0.831–0.800 = 0.31	0.873–0.817 = .056
Settings	Alpha NA Lambda = 0	Alpha = 1 Lambda = 0.146	Alpha = 0 Lambda = 1.000

Table 7.30 The coefficients for the L1 and L2 regularization logistic models

Variable	L1 coefficients	L2 coefficients	No regularization
Age		0.0054	−0.0267
Resting_BP		0.0024	0.0308
Cholesterol		0.0004	0.0034
Max_heart_rate_achieved	−0.0045	−0.0050	−0.0207
ST_depression	0.0646	0.1036	0.5118
male_Sex		0.2194	1.7348
typical_angina_Chestpain		−0.1659	−0.9120
atypical_angine_Chestpain		−0.1253	1.2608
aymptomatic_Chestpain	0.6114	0.3027	1.8305
FALSE_Fasting_sugar_gt_120mg		0.0279	0.9102
left_ventricular_hyperthophy_Resting_ECG		0.0906	0.6820
STT_T_abnormality_Resting_ECG		0.1509	1.0471
yes_Exercise_induced_angina		0.2390	0.5571
downsloping_Peak_exercise_ST		0.0053	−0.6939
flat_Peak_exercise_ST		0.1925	1.1594
One_Num_vessels_by_flourosopy		0.1981	2.3185
Two_Num_vessels_by_flourosopy		0.2840	3.1988
Three_Num_vessels_by_flourosopy		0.2966	3.0954
fixed_defect_Thalassemia		0.2090	1.1190
reversible_defect_Thalassemia	0.5363	0.3132	1.7544
Intercept	−0.0780	−0.9843	−6.8681

Interpreting the Coefficients

The results from the simpler L1 model will be used to interpret the coefficients. The baseline for comparison in this example set the predictors to the levels with the lower probability of heart disease. Each factor was then set in turn to its higher probability level. The difference from the baseline for each predictor indicates the variable's importance. Finally, all predictors were set at the levels resulting in the highest likelihood of disease. (Table 7.31).

The findings (Table 7.31) indicate that chest pain and a low maximum heart rate achieved in the tests were most important in predicting heart disease. A reversible Thalassemia defect (meaning a blood flow to the heart is observed, but it is not normal) is also a strong indicator of heart disease.

7.12   Asymmetric Benefits and Costs

The entire heart disease data set (training and test data) was submitted to the model created with L1 regularization. The confusion matrix resulting from this run of the logistic regression model is shown in Table 7.32. The accuracy metric differs slightly

Table 7.31 Effects of changes in predictors on the probability of heart disease

Settings	Asymptomatic chestpain	Max heart rate achieved	St depression	reversible defect thalassemia	Probability of heart disease	Difference from baseline
Baseline	No	202	0.0	No	0.271	0.000
Changes in probability due to changes in each predictor						
Chestpain	**Yes**	202	0.0	No	0.406	0.135
Max heart rate	No	**71**	0.0	No	0.402	0.131
ST depression	No	202	**6.1**	No	0.357	0.086
reversible_ defect	No	202	0.0	**Yes**	0.388	0.117

Table 7.32 Confusion matrix for heart data using the L1 logistic model

	Predicted heart disease	
Actual	Not present	Present
Not present	148	12
Present	27	110
Accuracy = (148 + 110)/(148 + 12 + 27 + 110) = .846		

from that in Table 7.29 since the entire data set was used rather than the training and test sets separately.

As is the case for many predictive problems, the consequences of an error can differ depending upon the type of error. In the heart disease case, the 27 cases where heart disease is present, but the model predicts not present are much more serious than the 12 cases predicting heart disease is present when it is not. The former error might lead to death or a serious heart attack, while the latter would cause stress for the patient. This might lead to further tests or monitoring, but generally, this is not as serious. Such situations have asymmetric costs (or benefits).

While putting a dollar value on the costs of false positives and false negatives might be possible in some contexts, another approach useful in medical situations is to create relative costs judgmentally by using a ratio. The judgmental ratio provides a way of incorporating risk into the model. Accurate predictions of either no heart disease or heart disease are assumed to be without cost. However, predicting that heart disease is present when it is not has a cost of one unit, while predicting that it is not present when it is present is assumed to cost 10 times as much. The following cost table (Table 7.33) shows the relative costs.

So, the cost ratio is 10:1 for the two costs. Using a ratio of 100:10 or any equivalent ratio will yield the same threshold value that results in the lowest cost. A higher ratio would force the model to predict heart disease "present" more frequently.

Table 7.32 shows that too many observations incorrectly predicted heart disease was not present. Lowering the threshold can increase the assignment to the heart

Table 7.33 Assumed costs
of errors

	Predicted heart disease	
Actual	not present	present
Not present	0	1 unit
Present	10 units	0

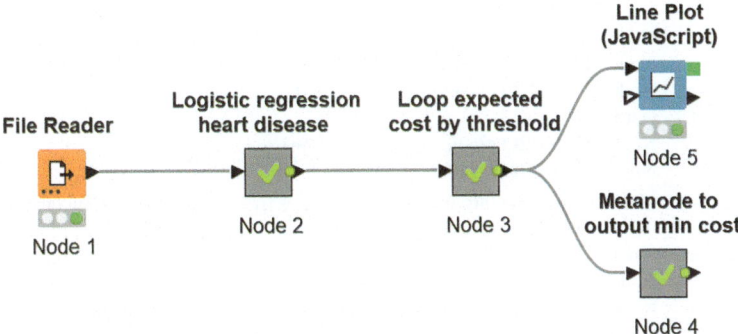

Fig. 7.21 Workflow to minimize cost with heart data

disease present outcome. A lower value for the threshold will reduce the cost. To assess the effects of changing the threshold, a loop was developed in KNIME, which systematically varying the threshold. Then the model selected the threshold value that minimized cost. The workflow in Fig. 7.21 does this. Node descriptions for the workflow are in Table 7.34. Figure 7.22 shows a chart of expected cost by threshold. The threshold that resulted in the lowest cost was found to be 0.32.

Tables 7.35 and 7.36 show the confusion matrices for the thresholds of 0.50 and 0.32, respectively. The expected cost was calculated by multiplying the costs from Table 7.33 by the cells of the confusion matrices. The cost using the threshold of 0.50 was found to be 282 units, while the cost with a threshold of 0.32 was 112 units.

While the lower threshold improved the model in terms of falsely predicting heart disease not present when it was present, the model would most likely not be acceptable to health professionals. The outcome of missing the detection of heart disease has been reduced to only a single case, but a large number of false positive results have been created. This tradeoff can be adjusted by changing the relative costs of the two errors. Still, the only way to achieve low false negatives *and* low false positives is to develop a better predictive model. This would require more observations, additional predictors, or a different, more accurate algorithm.

So far, the discussion has focused on making binary predictions. There are situations where the prediction levels are not binary. Strategies for evaluating multi-level outcomes were discussed earlier. An example of using KNIME to predict a multi-level outcome is shown in the next section.

Table 7.34 Node descriptions for the Fig. 7.21 workflow

Nodes	Descriptions	Input and output ports
Node 1 File reader	Read file: ClevelandHeartData.csv	<u>Output port:</u> data table with 297 rows and 14 columns
Node 2 Logistic regression heart disease	Meta node to prepare data and run logistic regression on the entire heart data set Node M2_1 one to many Node M2_2 column filter Node M2_3 H2O local context Node M2_4 table to H2O Node M2_5 H2O generalized linear model learner Node M2_6 H2O predictor (classification) Node M2_7 H2O to table	<u>Input port:</u> data table from the output port of node 1 <u>Output port:</u> data table with test data and predicted values of heart disease with 297 rows and 24 columns
Node 3 Loop expected cost by threshold	Metanode to loop through threshold values from 0.2 to 0.6 (41 rows) to compute misclassification "cost" Node M3_1 interval loop start Node M3_2 variable to table column Node M3_3 column expressions Node M3_4 math formula Node M3_5 row filter Node M3_6 loop end	<u>Input port:</u> data table from the output port of node 2 <u>Output port:</u> Estimated cost by threshold; 41 rows and 28 columns
Node 4 Metanode to output min cost	Sorts the cost results from lowest to highest; use row filter to only select top row with lowest cost; use column Filter to select only threshold and expected cost columns Node M4_1 sorter Node M4_2 row filter Node M4_3 column filter	<u>Input port:</u> data table from the output port of node 3 <u>Output port:</u> minimum cost
Node 5 Line plot	Create a line plot of expected cost versus threshold value Options: 　Column for x-axis: threshold_value 　Column for y-axis: Expected value	<u>Input port:</u> data table from the output port of node 3 <u>Upper output port:</u> Image of line plot <u>Lower output port:</u> Copy of input data

7.13 Multinomial Logistic Regression

What if the problem involves predicting more than two outcomes? For example, the target variable might be a political party: Democrats, Republicans, or Independents. Multinomial logistic regression can be used for targets with three or more categorical levels. The actual implementation is through decomposing the multi-class problem into multiple binary logistic regressions. The results are used to calculate the probabilities of each case belonging to different classes.

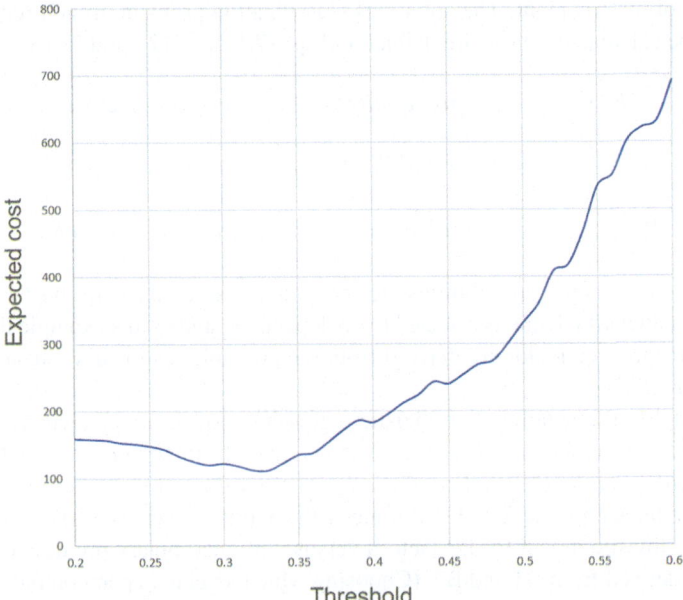

Fig. 7.22 The expected cost as a function of the threshold. The lowest cost was found at a threshold = 0.32

Table 7.35 Confusion matrix with a threshold = 0.50 (default)

| | Predicted heart disease | |
Actual	Not present	Present
Not present	148	12
Present	27	110
Cost = 27 × 10 + 12 × 1 = 282		

Table 7.36 Confusion matrix with a threshold = 0.32

| | Predicted heart disease | |
Actual	Not present	Present
Not present	58	102
Present	1	136
Accuracy = (58 + 136)/(58 + 102 + 1 + 136) = 0.65		

The multinomial model will be demonstrated using the famous Iris data set.[3] This data set contains 150 observations of three species of Iris flowers: Iris setosa, Iris virginica, and Iris versicolor, measured on four attributes: the length and the width of the sepals and the length and the width of the petals. The target variable is

---

[3] https://archive.ics.uci.edu/dataset/53/iris

the class of the Iris plant. The solve this classification problem, three probabilities need to be calculated. These are defined in Eqs. (7.11), (7.12), and (7.13).

$$P \mid y_{i=setosa} = the\ probability\ that\ the\ species\ is\ \text{Iris setosa} \tag{7.11}$$

$$P \mid y_{i=virginica} = the\ probability\ that\ the\ species\ is\ \text{Iris } virginica \tag{7.12}$$

$$P \mid y_{i=versicolor} = the\ probability\ that\ the\ species\ is\ \text{Iris } versicolor \tag{7.13}$$

Only two out of three probabilities are independent since the sum must be 1.0. It does not matter which one is selected to be dependent, and in this example the probabilities for Iris versicolor are derived from the probabilities for Iris setosa and Iris virginica.

To simplify the notation, two logistic regression models are created with different coefficients, as shown in Eqs. (7.14) and (7.15). The predictors are *sepal length$_i$*, *sepal width$_i$*, *petal length$_i$*, and *petal width$_i$*. B1 is used to estimate the setosa versus versicolor binary prediction. B2 estimates the virginica versus versicolor binary table. The logistic model for B3 (setosa versus virginica) binary prediction is algebraically derived from B1 and B2. (Choosing which level to depend on the others is arbitrary.)

$$let\,B_1 = b_{10} + b_{11} sepal\,length_i + b_{12} sepal\,width_i + b_{13} petal\,length_i$$
$$+ b_{14} petal\,width_i \tag{7.14}$$

$$let\,B_2 = b_{20} + b_{21} sepal\,length_i + b_{22} sepal\,width_i + b_{23} petal\,length_i$$
$$+ b_{24} petal\,width_i \tag{7.15}$$

The log odds are ratios of probabilities:

$$\ln\left(\frac{P\mid_{y_i=setosa}}{P\mid_{y_i=versicolor}}\right) = B_1 \tag{7.16}$$

$$\ln\left(\frac{P\mid_{y_i=virginica}}{P\mid_{y_i=versicolor}}\right) = B_2 \tag{7.17}$$

$$\ln\left(\frac{P\mid_{y_i=setosa}}{P\mid_{y_i=virginica}}\right) = B_1 - B_2 = \ln\left(\frac{P\mid_{y_i=setosa}}{P\mid_{y_i=versicolor}}\right) - \ln\left(\frac{P\mid_{y_i=virginica}}{P\mid_{y_i=versicolor}}\right) \tag{7.18}$$

Using Eqs. (7.16), (7.17), and (7.18), the three probabilities can be solved as shown in Eqs. (7.19), (7.20), and (7.21).

$$P \mid y_{i=setosa} = \frac{e^{B_1}}{1 + e^{B_1} + e^{B_2}} \qquad (7.19)$$

$$P \mid y_{i=verginica} = \frac{e^{B_2}}{1 + e^{B_1} + e^{B_2}} \qquad (7.20)$$

$$P \mid y_{i=versicolor} = \frac{1}{1 + e^{B_1} + e^{B_2}} \qquad (7.21)$$

In tools like KNIME, all the calculations are done automatically. The workflow for predicting the iris class is shown in Fig. 7.23. Node descriptions are in Table 7.37.

With this small data set, prediction is perfect (Table 7.38). The prediction accuracies and Cohen's Kappa are shown in Table 7.39.

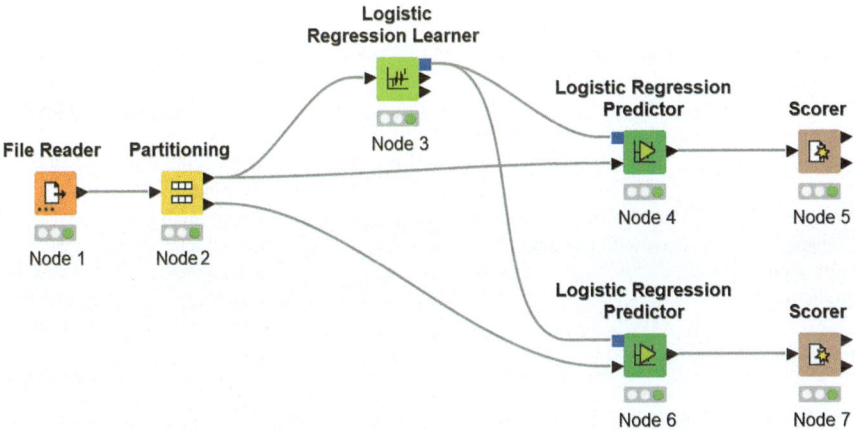

Fig. 7.23 Workflow for logistic regression with Iris data

Table 7.37 Node descriptions for the workflow in Fig. 7.23

Nodes	Descriptions	Input and output ports
Node 1 File reader	Read file: Iris.csv	Output port: data table with Iris data (150 × 5)
Node 2 Partitioning	First partition Select: relative: 80% Select: stratified sampling: Heart_Disease Check: use random seed: 123	Input port: data table from the output port of node 1 Upper output port: Iris training data (105 × 5) Lower output port: Iris test data (45 × 5)

(continued)

Table 7.37 (continued)

Nodes	Descriptions	Input and output ports

Node 3
Logistic
regression
learner
	Settings:
	    Target column: Iris class
	    Reference category: Iris-versacolor (as
	    discussed with a multinomial logit, one
	    level is selected as a reference)
	    Select solver: iteratively reweighted least
	    squares
	Feature selection:
	    All continuous predictor variables
		<u>Input port:</u> Data table from
		the upper output port of
		node 2
		<u>Upper output port:</u> logistic
		regression model
		<u>Middle output port:</u>
		coefficients and statistics
		<u>Lower output:</u> model and
		learning properties

Node 4 Logistic regression predictor	Use logistic regression model to predictor Iris class with training data	<u>Upper input port:</u> model from the upper output of node 3 <u>Lower input port:</u> data table from the upper output port of node 2 <u>Output port:</u> predictions of Iris class with training data (105×6)

Node 5
Scorer
	Create a confusion matrix and accuracy
	statistics with training data
	Scorer:
	    First column: Iris class
	    Second column: prediction (Iris class)
		<u>Input port:</u> data table from
		the output port of node 4
		<u>Upper output port:</u> confusion
		matrix
		<u>Lower output port:</u> accuracy
		statistics

Node 6 Logistic regression predictor	Use logistic regression model to predictor Iris class with test data	<u>Upper input port:</u> model from the upper output of node 3 <u>Lower input port:</u> data table from the upper output port of node 2 <u>Output port:</u> predictions of Iris class with test data (45×6)

Node 7
Scorer
	Create a confusion matrix and accuracy
	statistics with test data
	Scorer:
	    First column: Iris class
	    Second column: prediction (Iris class)
		<u>Input port:</u> data table from
		the output port of node 6
		<u>Upper output port:</u> confusion
		matrix
		<u>Lower output port:</u> accuracy
		statistics

Table 7.38 Confusion matrix for the test Iris data

		Predicted	
Actual	Setosa	Versicolor	Virginica
Setosa	14	0	0
Versicolor	0	19	0
Virginica	0	0	12

Table 7.39 Accuracy of
predicting Iris species in the
training and test data

	Training data	Test data
Accuracy	0.981	1.0
Kappa	0.957	1.0

7.14 Summary

Logistic regression is used to model and predict binary or categorical target variables, in contrast to linear regression which models continuous targets. The logistic model estimates the probability of an observation belonging to each outcome category, and these estimated probabilities are converted to class predictions using a threshold. Performance metrics like accuracy, precision, recall, and ROC curves are used to evaluate classification accuracy, considering tradeoffs between true and false positives/negatives. Cost ratios can be used to adjust the decision threshold when false positive and false negative consequences differ.

Logistic regression can be used with large numbers of observations; as with ordinary regression, the number of observations in an analysis should be much greater than the number of predictors. Also, care should be taken to avoid over-fitting (as with most supervised models).

Logistic regression assumes linearity between the log odds and the predictor variables, making interpreting the coefficients on predictors less intuitive than OLS regression. The use of odds ratios for interpretation was shown to be one approach. Another approach was to vary predictors one at a time to derive changes in the probabilities of a binary outcome.

Working with imbalanced data was demonstrated using the SMOTE node. When the costs of false positive and false negative errors differed, a process for adjusting the decision threshold was shown.

Appendix: Cohen's Kappa

Cohen's Kappa can be used to measure prediction accuracy with multi-level categorical target variables. Kappa is calculated by comparing the observed and expected values along the diagonal of a confusion matrix. The observed accuracy is simply the sum of the values in the table's diagonal divided by the total number of observations.

In the example in Fig. 7.24 the observed accuracy along the diagonal is $(22 + 27 + 54 + 47) = 150$. The expected values along the diagonal are computed, assuming that the row and column probabilities are independent. So, for example, the random expected accuracy for the cell in the upper left corner of the matrix (row1, column 1) in Fig. 7.24 can be estimated as the probability of an observation

4 X 4 confusion matrix

Actual	Predicted: Ideal	Premium	Good	Fair	
Ideal	22	7	10	18	57
Premium	5	27	7	6	45
Good	2	12	54	6	74
Fair	9	9	9	47	74
	38	55	80	77	250

Expected number in diagonal cells

Actual	Predicted: Ideal	Premium	Good	Fair	
Ideal	8.66				
Premium		9.9			
Good			23.68		
Fair				22.79	
					65.03

Fig. 7.24 Example for Cohen's Kappa calculation

Table 7.40 Interpreting Cohen's Kappa

Kappa value	Interpretation
0.80–1.00	Very good
0.60–.80	Good
0.40–.60	Moderate
0.20–.60	Fair
0.00–.20	Poor

being in row 1 times the probability of an observation being in column 1 times the total number of observations. The calculations are as follows:

$$\Pr\left(obs\,in\,row\,1\right)=\frac{57}{250}=0.228 \tag{7.22}$$

$$\Pr\left(obs\,in\,column\,1\right)=\frac{38}{250}=0.152 \tag{7.23}$$

$$\Pr\left(obs\,in\,cell\,1,1\right)=0.228\times 0.152=0.035 \tag{7.24}$$

$$Expected\left(obs\,in\,cell\,1,1\right)=0.035\times 250=8.66 \tag{7.25}$$

The calculations are repeated for each cell in the confusion matrix's diagonal. Equation (7.9) is repeated below in Eq. 7.26, substituting the actual and expected values along the diagonal substituted. Kappa is 0.45 (Table 7.27), which is moderate according to the commonly used interpretation shown in Table 7.40.

$$Cohen's\,Kappa=\frac{\left(observed\,diagonal\,obs-expected\,diagonal\,obs\right)}{\left(Total\,observations-expected\,diagonal\,obs\right)} \tag{7.26}$$

$$Cohen's\,Kappa=\frac{\left(150-65.03\right)}{\left(250-65.03\right)}=0.45 \tag{7.27}$$

References

Azen, R., & Traxel, N. (2009). Using dominance analysis to determine predictor importance in logistic regression. *Journal of Educational and Behavioral Statistics, 34*(3), 319–347.

Bex, T. (2021). *Comprehensive guide to multi-class classification metrics.* https://towardsdatascience.com/comprehensive-guide-on-multiclass-classification-metrics-af94cfb83fbd. Accessed 29 July 2023.

Kothawade, H. (2021). *Understanding receiver operating characteristic* (ROC) curve. https://medium.com/nerd-for-tech/understanding-receiver-operating-characteristic-roc-curve-f5eed11bc565. Accessed 29 July 2023.

Long, J. S., & Frees, J. (2006). *Regression models for categorical dependent variables using Stata* (2nd ed.). Chapman & Hall/CRC.

Provost, F., & Fawcett, T. (2013). Data science and its relationship to big data and data-driven decision making. *Big Data, 1.* https://www.liebertpub.com/doi/full/10.1089/big.2013.1508. Accessed 29 July 2023

Pykes, K. (2020). *Cohen's Kappa.* https://towardsdatascience.com/cohens-kappa-9786ceceab58. Accessed 29 July 2013.

Chapter 8
Classification and Regression Trees

Decision tree models (aka tree-based models) are commonly used in data mining to perform predictive analytics, typically with a single categorical or continuous dependent variable and multiple predictor variables (continuous or categorical). There are two principal decision tree types: classification and regression. The first section of this chapter discusses classification trees, and the second covers regression trees. Both types of decision trees are "automatic" in the sense that the independent variables are selected by searching for optimal splits using a measure of "purity" or "entropy."

8.1 Classification Trees

As an example, a classification tree could be used to inform customer targeting for a marketing campaign. Consider a hypothetical data set with the history of whether a customer purchased a product that was offered. The tree can identify the characteristics of past customers that were most likely to buy a product from the seller. Assume that 2500 observations are available with the following variables on each record:

- The target variable: Purchased the product or did not purchase the product.
- Predictor variables:

  - Income greater than $70 k or not.
  - Age over 30 or not.
  - College graduate or not.
  - Male or female.

Assume the data was analyzed with a classification tree with the results shown in Fig. 8.1. Each node in the diagram represents a set of customer observations. The

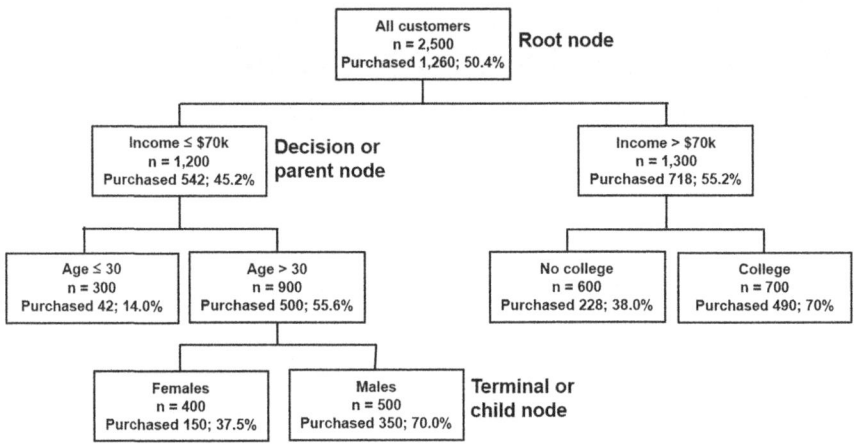

Fig. 8.1 A conceptual example of a classification tree

root node contains all 2500 customers in the database; 50.4% of the customers in this node purchased the product.

The algorithm continues by testing each predictor variable to determine which predictor splits the data into subgroups that are most pure (or homogeneous) with respect to the target variable. (Mechanisms for determining the splits include Gini impurity and entropy, which will be discussed later in this chapter.) The intermediate nodes are called decision or parent nodes, and the final nodes are called terminal or child nodes.

The diagram shows that the first split is by income. Customers with incomes above $70 k had a purchase rate of 55.2%, while those with lower incomes had a purchase rate of 45.2%. The algorithm further splits the lower-income group by age and gender, while the higher-income group is divided by education.

The variables selected for splitting at each level need not be the same. If the tree branches do not result in parallel splits, this indicates an interaction effect between the predictor variables. This example shows an interaction of income with age and education. If income is low, then age is the next significant predictor, whereas if income is high, the next split is by education. The decision rules from this hypothetical example are:

- If Income ≤ $70,000 AND Age ≤ 30, THEN the purchase probability = 14.0%.
- If Income ≤ $70,000 AND Age > 30 AND Female, THEN the purchase probability = 37.5%.
- If Income ≤ $70,000 AND Age > 30 AND Male, THEN the probability of purchase = 70.0%.
- If Income > $70,000 AND No college, THEN the probability of purchase = 38.0%.
- If Income > $70,000 AND College graduate, THEN purchase probability = 70.0%.

Two promising target segments are identified in this hypothetical example: Males over 30 with lower incomes and college graduates with higher incomes. Both had purchase rates of 70.0%, suggesting that these groups would be attractive target groups for additional purchases.

8.2 Applications of Decision Trees

Tree-based models have been around for a long time, and until recent years, some researchers warned against using decision trees. One of the first models, Automatic Interaction Detection (Sonquist & Morgan, 1964), was derisively called a "substitute for thinking." By automatically combing through data sets in search of relationships, tree models can "discover" spurious associations that may appear plausible but are only artifacts due to randomness or idiosyncratic effects in a specific data set. That is especially true when decision trees are used on small data samples. Even with large data sets, it is critical that training and testing subsets or k-fold partitioning methods are used to assess the ability of the models to predict unseen data.

Despite the limitations noted above, decision trees can be used to create predictive models in situations without a well-defined theory. Some of the common application areas of decision trees are:

- *Prediction* – creating rules and using them to predict future outcomes.
- *Market segmentation* – identifying segments most likely to purchase.
- *Data reduction and variable screening* – screening many variables to determine the best prospects.
- *Interaction detection* – finding variables with effects that differ according to levels of other variables.
- *Category merging* – recoding variables with many categories into fewer categories without substantial loss of predictive information.

There are two essential phases to consider in analyses with decision trees: (1) developing the tree and (2) pruning the tree.

8.3 Developing Classification Trees

There are exponentially many possible trees with a given set of predictors since every value can split predictors, and the same predictor can be used repeatedly as the tree is built. Since examining all possible tree structures in each problem is usually impossible, an algorithm is used to grow a reasonably accurate tree instead of the optimal tree. Decision trees use so-called "greedy algorithms" to form splits in a sequential process. Once a division in the data is made, the algorithm does not go back after further divisions are made to check that previous splits were optimal.

Thus, at each stage, the split selections are locally optimal. This yields a tree that is usually a very good model but not globally optimal.

Various algorithms have been developed, all of which fall under the umbrella term of classification trees. Tree algorithms differ in terms of:

- The criterion used to determine how to split nodes (e.g., Gini impurity, entropy).
- Whether splits are only binary or can be multi-level.
- Stopping rules (e.g., the number of splits, the minimum size node that can be split).
- The types of variables can be used; most algorithms can handle a mix of categorical and continuous variables.
- The output structure and level of detail shown.

Three of the more popular methods for classification and regression trees are:

- CART (Classification and Regression Trees by Breiman et al., 1984)

    - Gini impurity is the measure used for evaluating splits.
    - Input variables can be categorical or continuous.
    - Splits are binary.

- C50 (an updated version of the model C4.5 by Quinlan, 1993)

    - Entropy is used to determine splits.
    - Input variables can be categorical or continuous.
    - Splits can be binary or multiway.

- CHAID (Chi-square Interaction Detection by Kass, 1980)

    - A chi-square test is used to determine splits.
    - Input variables can be categorical or continuous.
    - Splits can be binary or multiway.

For continuous predictor variables, splits are evaluated at each variable value. Thus, for n distinct values of a predictor, n-1 potential partitions are considered. For nominal predictor variables, the number of divisions into two groups depends on the number of distinct categories. Table 8.1 gives examples of splits as a function of distinct categories in a nominal variable. In general, the number of splits as a function of categories is given by Sterling Numbers of the Second Kind (Weisstein, n.d.).

Table 8.1 Possible binary splits by the number of categories in a nominal predictor

Categories	# possible splits
2	1
3	3
4	7
5	15
6	31
7	63

8.4 Growing Decision Trees Using Gini Impurity

Developing a classification tree with a binary target is a sequential process of sub-dividing parent nodes into child nodes to create pure nodes in terms of the target variable. Two criteria that can be used to identify the best splits in a tree are Gini impurity and information gain based on changes in entropy (Aznar, 2020).

The process of building a tree will be illustrated using the Gini criteria. The process using entropy (Shannon, 1948) is quite similar. For Gini impurity purity measure G_{Parent}, is computed for a parent node.

The Gini impurity, G_{Parent}, of the parent node in a classification tree with a binary target is computed as:

$$G_{Parent} = 1 - \left(p_0^2 + p_1^2 \right) \tag{8.1}$$

where p_0 is the proportion of cases in a node at level "0" and
p_1 is the proportion of cases in a node at level "1".

Figure 8.2 shows the Gini impurity as p_0 and p_1 are varied. The Gini impurity is zero when the node has only zeros or only ones. The highest value of Gini impurity occurs when there is an equal number of ones and zeros and lowest when the node contains either all 0's or all 1's.

Demonstrating Gini Calculations

The process for growing the tree is illustrated in Fig. 8.3.

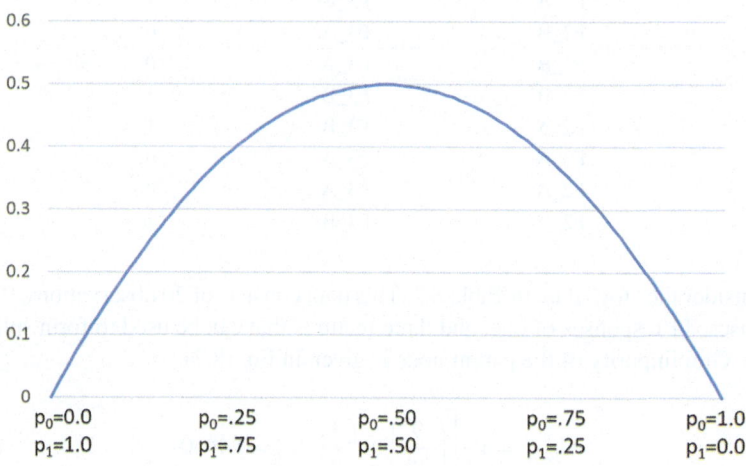

Fig. 8.2 The Gini impurity as a function of p_0 and p_1

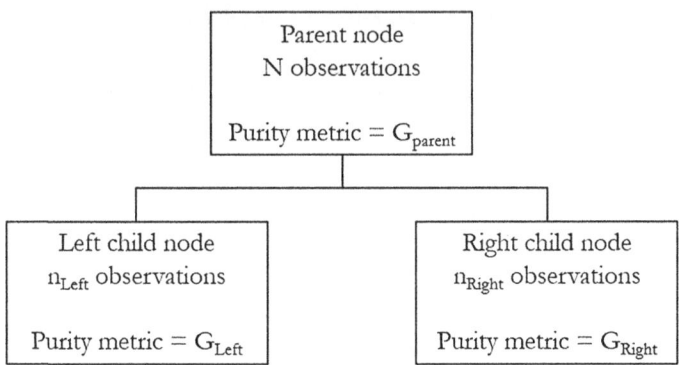

Fig. 8.3 Purity metrics for splitting a node

Table 8.2 A 'toy' data set to illustrate a decision tree

Feature 1	Feature 2	Feature 3	Response
F1_A	F2_A	F3_A	0
F1_A	F2_A	F3_B	1
F1_A	F2_A	F3_A	1
F1_B	F2_B	F3_A	1
F1_B	F2_B	F3_B	1
F1_B	F2_B	F3_B	0
F1_A	F2_B	F3_B	1
F1_A	F2_B	F3_A	0
F1_A	F2_B	F3_B	1
F1_B	F2_B	F3_B	1
F1_A	F2_B	F3_A	0
F1_A	F2_B	F3_A	0
F1_A	F2_A	F3_B	1
F1_B	F2_B	F3_A	0
F1_B	F2_B	F3_A	0
F1_B	F2_B	F3_B	1
F1_A	F2_A	F3_B	1
F1_A	F2_A	F3_A	0
F1_A	F2_A	F3_A	0
F1_A	F2_A	F3_B	1

Consider the "toy" data in Table 8.2. This data consists of 20 observations, 9 zero responses, 11 responses of one, and three features that can be used to form splits. The Gini impurity of the parent node is given in Eq. (8.2).

$$G_{Parent} = 1 - \left[\left(\frac{9}{20} \right)^2 + \left(\frac{11}{20} \right)^2 \right] = 0.4950 \tag{8.2}$$

The Gini calculations for splitting by F1, F2, and F3 are in Figs. 8.4, 8.5, and 8.6. The reduction in Gini impurity is computed for each possible split and given by Eqs. 8.3, 8.4, and 8.5:

$$G_{Reduction\ for\ split\ on\ F1} = G_{Parent} - G_{split\ on\ F1} = 0.4950 - 0.4945 = 0.0005 \quad (8.3)$$

$$G_{Reduction\ for\ split\ on\ F2} = G_{Parent} - G_{split\ on\ F2} = 0.4950 - 0.4542 = 0.0408 \quad (8.4)$$

$$G_{Reduction\ for\ split\ on\ F3} = G_{Parent} - G_{split\ on\ F3} = 0.4950 - 0.2500 = 0.2450 \quad (8.5)$$

The largest reduction in impurity is obtained by splitting the parent node by F3. Next, the reduction in Gini impurity by splitting each of the four nodes in the revised tree: F1, F2, F3A, and F3B will be determined. The process will continue until no further splits are possible.

Feature 1_A		Feature 1_B	
Response	Cases	Response	Cases
0	6	0	3
1	7	1	4

$$Gini_{1\_A} = 1 - \left(\frac{7}{13}\right)^2 - \left(\frac{6}{13}\right)^2 = 0.4970$$

$$Gini_{1\_B} = 1 - \left(\frac{3}{7}\right)^2 - \left(\frac{4}{7}\right)^2 = 0.4898$$

$$Gini_1 = 0.4970 \times \frac{13}{20} + 0.4898 \times \frac{7}{20} = 0.4945$$

Fig. 8.4 Gini calculations for splitting on Feature 1

Feature 2_A		Feature 2_B	
Response	Cases	Response	Cases
0	3	0	4
1	5	1	8

$$Gini_{2\_A} = 1 - \left(\frac{3}{8}\right)^2 - \left(\frac{5}{8}\right)^2 = 0.4688$$

$$Gini_{2\_B} = 1 - \left(\frac{4}{12}\right)^2 - \left(\frac{8}{12}\right)^2 = 0.4444$$

$$Gini_2 = 0.4688 \times \frac{8}{20} + 0.4444 \times \frac{8}{20} = 0.4542$$

Fig. 8.5 Gini calculations for splitting on Feature 2

Feature 3_A	
Response	Cases
0	8
1	2

Feature 3_B	
Response	Cases
0	1
1	9

$$Gini_{3\_A} = 1 - \left(\frac{8}{10}\right)^2 - \left(\frac{2}{10}\right)^2 = 0.3200$$

$$Gini_{3\_B} = 1 - \left(\frac{1}{10}\right)^2 - \left(\frac{9}{10}\right)^2 = 0.1800|$$

$$Gini_3 = 0.3200 \times \frac{10}{20} + 0.1800 \times \frac{10}{20} = 0.2500$$

Fig. 8.6 Gini calculations for splitting on Feature 3

8.5 Pruning to Avoid Overfitting

Splitting nodes can be continued until all nodes are pure (100% zeros or ones) or when no features are available for further splits. In actual applications, large data sets with thousands of observations and dozens of predictors can lead to huge trees with hundreds of nodes. Allowing the tree to grow without constraint can lead to overfitting of the tree to the data used (Bramer (2007). Overfitting occurs when the classification tree performs well on training data but poorly on test data. Some degree of overfitting will almost always occur because the training data and the test do not contain the same pattern of instances.

As an example of the effect of overfitting, a data set with over 13,000 observations, 23 predictor variables, and a binary target was analyzed using the R package rpart. The data was partitioned into 50% training and 50% test subsets. A sequence of decision trees was created that varied from one to over 1300 splits, and the prediction accuracy was recorded for the training and test subsets was recorded. Figure 8.11 shows the accuracy as a function of the number of splits (or size) of the tree. Accuracy with the training data and test data are approximately the same until the number of splits exceeds about 32. With further splits, the accuracy observed in the training data continues to increase, eventually approaching 1.0. This is evidence of overfitting since the accuracy of the test data falls off.

To avoid overfitting situations such as that shown in Fig. 8.7, pruning removes some of the tree's lower branches. The premise is that many of the low branches are not likely to increase the model's predictive accuracy with new data and may lead to overfitting. Two general approaches to pruning are available:

- Pre-pruning – early stopping of the tree-building process before all splits are done.
- Post-pruning – the splitting process can continue until no further splits are possible, and then removing branches to simplify the tree.

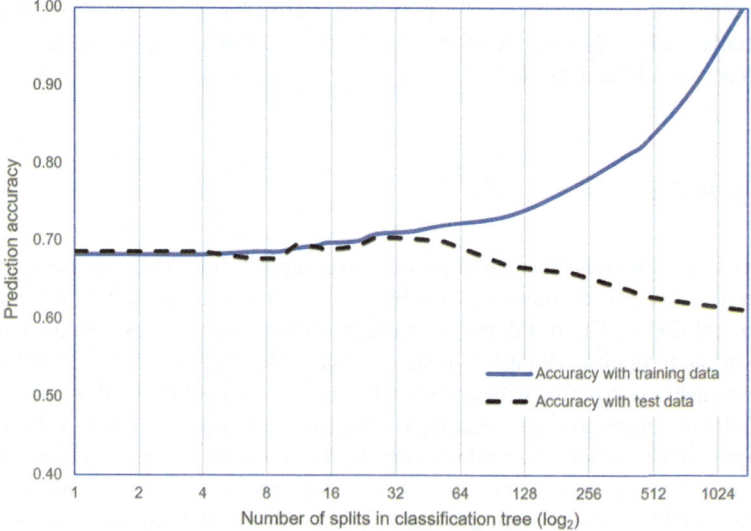

Fig. 8.7 Prediction accuracy as a function of number of splits

Pre-pruning

Pre-pruning avoids growing large trees. This can be accomplished by setting stopping rules for the tree algorithms. Typical stopping rules for pre-pruning are based on the depth of the tree, the minimum number of observations in nodes to split, or the minimum size of the terminal nodes. Three pre-pruning rules are:

1. Set a maximum depth of the tree, measured by the number of levels from the root node to the farthest terminal node.
2. Set a minimum number of observations in a node considered for splitting. For example, if the minimum number of records is set to 10, only nodes with 10 or more records can be split.
3. Set a minimum number of observations that must be present in any terminal node. For example, if the minimum number is set to 10, then every terminal node (leaf node) must have at least 10 records. (If this minimum is set too low (e.g., 1 or 2), overfitting is likely since this does not constrain the splitting process very much.)

While pre-pruning is easy to implement, it is difficult to determine the proper settings for the threshold for stopping tree growth. If the stopping threshold is too conservative (e.g., by allowing only a small number of branches), the resulting tree will not be as accurate as it might be. On the other hand, too liberal a threshold can lead to overfitting. Bramer (2007) investigated prediction accuracy on test data for 11 different data sets using various settings of pre-pruning parameters. He concluded that "… the choice of pre-pruning method is important. However, it is essentially ad hoc. No choice of size or depth cutoff consistently produces good results across all the datasets."

Post-pruning

With post-pruning (aka "pruning upward"), a complete, full-depth tree is generated and then trimmed by eliminating some branches, converting selected decision nodes to terminal nodes. The full depth is attained when terminal nodes contain a single observation, have observations with the same target value, or no improvement in the purity measure is possible with further splitting. Creating a full-depth tree can consume a lot of computing time and create a huge tree. Therefore, a combined pre-and post-approach is sometimes used wherein the tree-growing process is stopped when the terminal nodes have a minimum number of observations. For example, pre-pruning could be used by setting at least five records in any terminal node and then using a post-pruning to reduce the tree size further.

Despite the lack of one best method, post-pruning remains a valuable approach to avoid overfitting classification trees. Judgment and experimentation with different techniques may be helpful with specific data sets. To that end, three post-pruning methods will be described that are available in KNIME: Reduced Error Pruning (REP), Cost-Complexity Pruning (CCP), and Minimum Description Length Pruning (MDL).

Reduced Error Pruning

Reduced error pruning (REP) divides the data in a training set into a "growing set," one or more "pruning set(s)," and a test set to use as a final check. (Fig. 8.8). The growing set is used to create the full tree.

Then, using a bottom-up approach, each decision node is temporarily collapsed into a terminal node, and the misclassification rate is calculated using the pruning set. The pruning set is unseen data at this point. If the error with the pruning set is reduced or the same, the decision node under consideration is turned into a child node. The process continues until further pruning increases misclassification errors. A final assessment of error can then be made using the test set.

Fig. 8.8 Data partitions for reduced error post-pruning

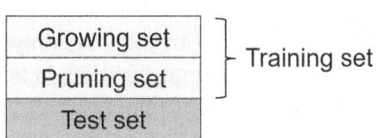

Reduced error pruning is quite simple and works well when a large number of observations are available for creating the training, pruning, and test sets. Esposito and Malerba (1997) and Quinlan (1987) noted that reduced error pruning could be biased toward over-pruning, especially in cases where the pruning set is much smaller than the training set. In such cases, rarer exceptional cases in the entire data set may not be represented in the test set.

Cost-Complexity Pruning

Cost-complexity pruning was developed by Breiman et al. (1984) as part of the CART algorithm. As the number of splits increases in a tree, the prediction error will decrease or remain the same. To find a "good" tree, the algorithm proceeds by growing a much too large tree and then pruning from the bottom up. The following expression is used to calculate the Complexity Parameter C_P for each sequential partition of subtrees:

$$C_P = \frac{\text{Error after pruning} - \text{Error prior to pruning}}{\text{\# nodes prior to pruning} - \text{\# nodes after pruning}} \tag{8.6}$$

The cost-complexity pruning method proceeds as follows:

1. A sequence of trees is created with the number of splits ranging from a fully grown tree down to the root node.
2. The prediction error is calculated for each tree in the sequence, and the number of splits is recorded.
3. The C_P value is computed, which compares the increase in splits and reduction in error for the sequence of trees as more splits are done. The tree with the minimal C_P value is selected for further consideration for each given number of splits.
4. The misclassification rate for each tree selected in step 3 is computed using a k-fold validation with the training data set or, if the number of observations is sufficiently large, with a separate hold-out validation sample.

Since the pruned tree with the lowest C_P is selected for each number of splits, the number of subtrees to be evaluated in step 4 will usually be much less than the total number of possible subtrees. The process of computing C_P is illustrated in Figs. 8.9 and 8.10. Figure 8.9 shows a hypothetical tree with six splits and seven child nodes (indicated by C). The internal nodes are indicated by D. Figure 8.10 shows the tree with two splits pruned. Note that the node above the removed splits is changed from a decision node to a child node. In doing so, the prediction error will increase (or remain the same). However, the difference between the two splits in the denominator reduces C_P.

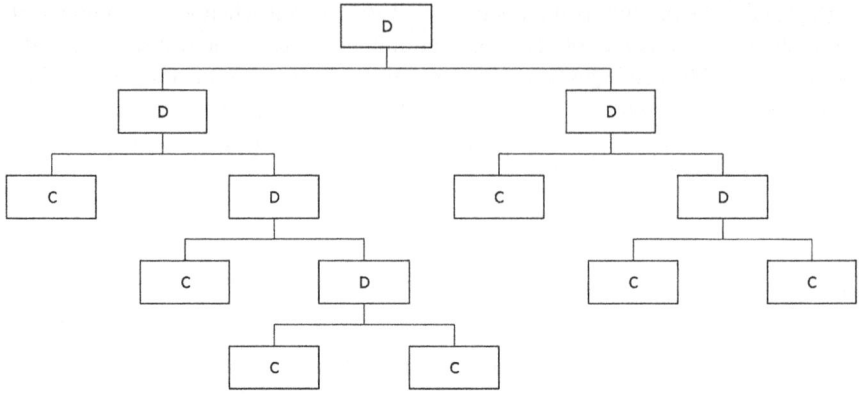

Fig. 8.9 The structure of a full tree

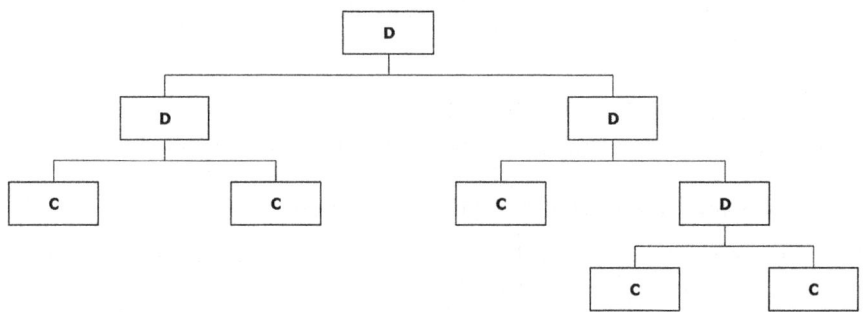

Fig. 8.10 The structure of a pruned tree

The complexity parameter (C_P) for the pruning of the tree in Fig. 8.9 to that in Fig. 8.10 is calculated as follows, with $C_p = (ET_2 - ET_1)/(6-4)$ with results shown in Table 8.3.

Pruning Using the Minimum Description Length Principle

The Minimum Description Length principle (MDL) measures the complexity of a classification tree using concepts from information theory. This principle states that the best tree is the one that minimizes the cost of information transmission measured in bits to encode the model. Several different approaches to MDL pruning have been developed, e.g., Quinlan and Rivest (1989), Wallace & Patrick (1993), and (Kononenko (1998).

Table 8.3 Number of splits by prediction errors

Tree	Number of splits	Prediction error
Figure 8.9	6	ET_1
Figure 8.10	4	ET_2

Table 8.4 Pruning methods available in KNIME

Method	KNIME node
Pre-pruning	
Minimum number of records per node	Decision tree learner
Minimum number of records per node	REPTree in the Weka 3.7 integration
Post-pruning	
Reduced error pruning	REPTree in the Weka 3.7 integration
Cost-complexity pruning in CART	rpart in R integration
Minimum description length pruning	Decision tree learner

The MDL process starts with growing a full tree and evaluating each split from the bottom up. No hold-out sample is used, which is an advantage of this pruning technique. The cost of information transmission is the sum of the information cost of the tree itself plus the cost of misclassification errors. The cost of the tree grows as the branches are added. The cost of classification errors shrinks as the model's predictive error decreases. The model is given by Eq. (8.7):

$$MDL = L(model) + L(data \mid model) \qquad (8.7)$$

where L(model) = the cost in information bits to encode the tree model and L(data|model) = the cost in information bits to encode the misclassified instances.

Recommendations on Pruning

No single pruning method is the best to use with decision trees. However, some approach to pruning must be used to avoid creating models that are unlikely to generalize. In addition, while some pruning techniques do not require a separate testing or validation data set, it is still a good practice to use unseen data as a final check on predictive accuracy.

There are several pruning methods available in KNIME (Table 8.4).

Different pruning approaches will yield different decision trees, so the final judgment on which method to use is up to the analyst. If the number of observations is sufficiently large, then it is strongly recommended that a separate hold-out sample is retained for the final verification of prediction accuracy. Remember that decision trees, as greedy algorithms, are built from the top down, and there is no guarantee that a tree is optimal, even after pruning. The criterion is one of pragmatism: is the accuracy of the tree sufficient for the problem?

8.6   Missing Values in Decision Tree Analyses

Since missing values for some attributes are likely to occur in real data sets, the question of how classification trees handle such cases arises. Despite claims to the contrary, decision trees are not immune to the effects of missing values, and different approaches to managing continuous versus nominal variables are needed. For nominal predictors, missingness is sometimes handled by assigning missing values as simply another variable level. However, Pyle (1999) reports this is not always satisfactory since assigning all missing values to the same category assumes all such cases have the same value.

It is usually better to deal with missing values as part of the data cleaning process rather than rely on the decision tree algorithm to handle such cases. However, some decision tree algorithms have explicitly incorporated one or more ways of handling instances with missing information. Missing values can occur in training, validation, or new data submitted to the model. Twala (2009) considered three ways to handle missing data in decision trees: (1) ignoring missing values; (2) imputation; and (3) using machine learning.

Ignoring Missing Data

While the most straightforward approach is to drop cases with missing values, this is usually not a good practice. This approach might be justified if the proportion of cases with missing values is small. For example, it should be noted that some decision tree software programs adopt this method of simply dropping cases with missing values. This is the case with the Decision Tree Learner in KNIME.

Imputation Techniques

Imputation methods employ information in a training set to replace missing values. The replacement value could be the mean, median, or mode for continuous attributes. A disadvantage of this approach is that the distribution of variables can be distorted, and the variance can be reduced. For nominal attributes, the replacement value could be the most common level of the attributes. However, with nominal (categorical) attributes, viewing missing as another class may be better.

Regression may be used to impute missing values with continuous variables. Let X = the predictor variable with missing values. Temporarily remove all observations where X is missing. Next, regress X as the target on all the remaining predictor variables. The resulting regression model is used to estimate the missing X values, which are then substituted for the missing values. This approach is repeated for each variable that has missing values. This approach has the advantage over simply using a single value, such as the mean or median, which implicitly assumes that all observations with missing values have the same value. Predictive models such as logistic regression or decision trees can be used analogously for nominal missing values.

Using Machine Learning

CART and its implementation in *rpart* use an approach called surrogate splits. To use the surrogate splits approach, consider a situation with a set of predictor attributes A_i, $i = 1$ to k. Assume that the specific attribute A_j has several observations with missing values. When evaluating a split using attribute A_j at a specific parent node into child nodes, only those instances of A_j with known values are used to compute the purity of the left and right child nodes. Once this is done, the next step is assigning each observation to the left or right child nodes. This presents a problem for instances that need to include values on attribute A_j. A surrogate splitting variable most highly correlated with A_j is then selected from the set A_i. For those cases where A_j is missing, the surrogate attribute is used to make an assignment to the left or right child node for that case.

8.7 Outliers in Classification Trees

Classification trees are said to be robust to outliers in continuous predictors. Splits on predictors that contain either positive or negative outliers will not influence the resulting tree since the value selected for the split is usually far from the extremes. (However, outliers can distort regression trees.)

It is commonly believed that classification trees are not affected by monotone transformations on the predictor variables since the order of the values is not changed. This idea means that transformations to reduce skewness or feature scaling, such as normalization or standardization of predictors, are unnecessary and will have little or no effect on model performance. While the general rule is valid, there are certain conditions where transformations can affect accuracy (Galili & Meilijson, 2020).

8.8 Predicting Churn with a Classification Tree

This example uses a popular data set on customer churn in a telecom company. Churn occurs when a customer (player, subscriber, user, etc.) ceases their relationship with a company. The total cost of customer churn includes lost revenue and the marketing costs of replacing those customers with new ones. Reducing customer churn is a key business goal of every online business. It is almost always more difficult and expensive to acquire a new customer than to retain a current paying customer. The ability to predict which customers are at a high risk of churning while there is still time to do something about it represents a significant potential for increasing revenue and profit.

In the case of telecom companies, customers may cancel for many reasons, including poor service, availability of specific hardware, and price. Identifying potential churners before they quit can be the first step to reducing the churn rate. It makes financial sense to offer incentives only to potential churners and not customers who are predicted to remain with the company. Analytic techniques can be used to develop predictive models to identify likely churners based on customer characteristics and behaviors.

For this example, a data set, Telecom Customer Churn Prediction,[1] consisting of 7043 telecom provider customers, is available with the variables shown in Table 8.5.

A classification tree was developed in KNIME using the workflow shown in Fig. 8.11. The nodes in the workflow are described in Table 8.6.

Table 8.5 Variables in the TelcoChurn5000.csv data set

Variable	Description
gender	Male or female
SeniorCitizen	0 or 1
Partner	Yes or no
Dependents	Yes or no
tenure	# months with carrier
PhoneService	Yes or no
MultipleLines	Yes, no or no phone
InternetService	DSL, fiber optic, no
OnlineSecurity	Yes, no, no internet service
OnlineBackup	Yes, no, no internet service
DeviceProtection	Yes, no, no internet service
TechSupport	Yes, no, no internet service
StreamingTV	Yes, no, no internet service
StreamingMovies	Yes, no, no internet service
Contract	Month-month, 1 year, 2 year
PaperlessBilling	Yes or no
PaymentMethod	Bank transfer (automatic), credit card(automatic), electronic check, mailed check
MonthlyCharges	Dollars per month
TotalCharges	Total dollars during tenure
Churn	Yes or no

[1] https://community.ibm.com/community/user/businessanalytics/blogs/steven-macko/2019/07/11/telco-customer-churn-1113

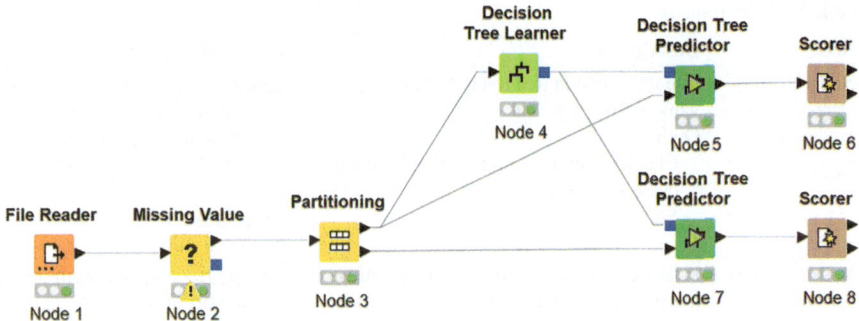

Fig. 8.11 KNIME workflow for a classification tree using churn data

Table 8.6 Node descriptions for the Fig. 8.11 workflow

Nodes	Descriptions	Input and output ports
Node 1 File reader	Read file: ChurnData.csv.	<u>Output port:</u> file table with 7043 rows and 20 columns
Node 2 Missing value	Default: impute mean values for 11 missing values on total charges	**Input port:** data table from output port of node 1. <u>Output port:</u> file table with 7043 rows and 20 columns.

Node 3
Partitioning
	First partition:
	    Training select relative: 60%
	    Select stratified sampling: churn
	    Check: use random seed: 123
		<u>Input port:</u> data table from the output port of node 2
		<u>Upper output port:</u> training data with 4225 rows and 20 columns
		<u>Lower output port:</u> test data 2818 rows and 20 columns

Node 4
Decision tree learner
	This node produces a classification tree
	Options:
	    Class column: churn
	    Quality measure: Gini Index
	    Pruning method: MDL (minimum description length) and reduced error pruning
	    Min number of records per node: 50
		<u>Input port:</u> data table from the upper output port of node 3
		<u>Output port:</u> decision tree model

Node 5
Decision tree predictor
	Use the decision tree to predict the class value for the training data
	Options:
	    Check append columns with normalized class distribution
	    Suffix = Prob
		<u>Upper input port:</u> model from the output port of node 4
		<u>Lower input port:</u> data table from the upper output port of node 3
		<u>Output port:</u> training data plus predicted values

Node 6
Scorer
	Produces accuracy statistics and a confusion matrix for the decision tree model training set
	Scorer:
	    First column: churn
	    Second column: prediction (churn)
		<u>Input port:</u> data table from the output port of node 5
		<u>Upper output port:</u> confusion matrix
		<u>Lower output port:</u> accuracy statistics

(continued)

Table 8.6 (continued)

Nodes	Descriptions	Input and output ports
Node 7 Decision tree predictor	Use the decision tree to predict the class value for the test data Options: Check append columns with normalized class distribution Suffix = Prob	Upper input port: model from the output port of node 4 Lower input port: data table from the lower output port of node 3 Output port: test data plus predicted values
Node 8 Scorer	Produces accuracy statistics and a confusion matrix for the decision tree model test set Input port: test data plus predicted values Scorer: First column: churn Second column: prediction (churn)	Input port: data table from the output port of node 7 Upper output port: confusion matrix Lower output port: accuracy statistics

Table 8.7 Performance measures for the classification tree

Model	# splits	Accuracy with training data	Accuracy with test data
No pruning	Hundreds	0.997	0.721
Pruning	11	0.813	0.789

The analysis was rerun without any pruning and the number of nodes, the Training set accuracy, and the Test set accuracy were recorded, The results (Table 8.7) show that the pruning greatly reduced the number of nodes in the tree while slightly increasing the Test set's accuracy.

8.9   Regression Trees

Decision trees can also be used to predict continuous target variables. In such applications, these are known as regression trees. The analysis is much like the case with categorical target variables: the model produces a series of splits using selected predictor variables. Much like ordinary regression, you can have continuous or nominal predictors with regression trees. However, the other assumptions associated with linear regression regarding error distributions and so on are irrelevant because regression trees are not a classical statistical technique.

Regression trees operate by successively dividing a data set into smaller and smaller groups that are more homogeneous with respect to the target variable. The groups are nodes, and each node lower in the tree is more homogeneous in the target variable than those higher in the tree. The model starts with no predictors and then examines each predictor to select the best variable for the initial split.

The process of building a regression tree is illustrated using a simple data set consisting of 100 observations of home prices as the continuous target variable with

Table 8.8 Demo file for regression tree

Row	Price in $000's	Area in sq. ft.	Quality
1	203.8	1000	High
2	195.0	1000	Average
3	200.9	1000	High
4	403.2	2000	High
5	402.8	2000	High
...	...	...	...
96	306.9	2000	High
97	194.3	1000	Average
98	206.2	1000	High
99	310.1	2000	High
100	298.0	2000	Average

area in square feet (either 1000 or 2000) and quality (either high or average) as the predictors. The first five and last five rows are shown in Table 8.8.

The tree-building process uses the criterion of minimizing the sum of squared errors as binary splits in the predictors are made. The initial sum of squares (SST_I) is given by Eq. 8.8:

$$SST_I = \sum_n^{i=1}\left(y_i - \bar{y}\right)^2 \tag{8.8}$$

where n = the number of observations,
y_i = the price for observation i, and
$\bar{y}$ = the mean value of the prices.

Next, splits are considered by area in square feet and quality. When a split is done, each split branch's sum of squares (SST) is computed and added together. After splitting, the SST is subtracted from SST_I, and the split with the greatest reduction in SST is selected.

The sum of squares for each binary split is given by Eq. (8.9):

$$SST_{binary} = \sum_{n_L}^{i=1}\left(y_i - \bar{y}_L\right)^2 + \sum_{n_R}^{i=1}\left(y_i - \bar{y}_R\right)^2 \tag{8.9}$$

n_L = the number of observations in the left node after the split.
n_R = the number of observations in the right node after the split.

The results of the calculations are shown in Table 8.9. Splitting on Area produced the largest reduction in the sum of squares, which indicates that the first split should be with Area.

A recursive process is used in a larger problem, and each predictor is considered. Each potential split of a continuous predictor must be calculated so that for a variable with k distinct values, k-1 splits are considered.

Table 8.9 Calculations on
the first split of the
regression tree

Source
Initial total SSE = 502,301
Split on quality
Total SSE in Price for average quality node = 78,014
Total SSE in Price for high-quality node = 397,636
Total SSE after the split on quality = 475,649
Reduction in SSE for the split on quality = 26,651
Split on area
Total SSE in Price for 1000 sq. ft. node = 1226
Total SSE in Price for 2000 sq. ft. node = 98,421
Total SSE after the split on area = 99,647
Reduction in SSE for a split on area = 402,654

There are several different algorithms for regression trees, but a commonly used approach is the CART method (Breiman et al., 1984). The programs can control the depth of the tree. Since the target variable is continuous, the predicted value in each terminal node equals the average value of records in the node. Since regression trees have a finite number of terminal nodes, the predicted values will have a limited number of distinct values (cardinality). This fact can reduce the predictive accuracy when using regression trees compared with ordinary linear regression.

8.10   Example: Head Acceleration in a Simulated Motorcycle Accident

This example uses a data set, mcycle.csv, from the R package MASS which provides a series of measurements of head acceleration in a simulation used to test crash helmets. The simulations are typically carried out with a test dummy. The impact absorption capacity of a helmet is determined by recording the acceleration versus time imparted to a head-form fitted with helmet when dropped in guided free fall at a specific impact velocity upon a fixed steel anvil. The data set has 133 observations of time (in milliseconds) and acceleration.

Figure 8.12 shows a plot of the actual and predicted accelerations. The predictions were made using a regression tree in KNIME. The workflow is in Fig. 8.13 and the node descriptions are in Table 8.10.

The regression tree smoothed the relationship between acceleration and time. Ordinary regression would not easily capture the non-linearity of the data.

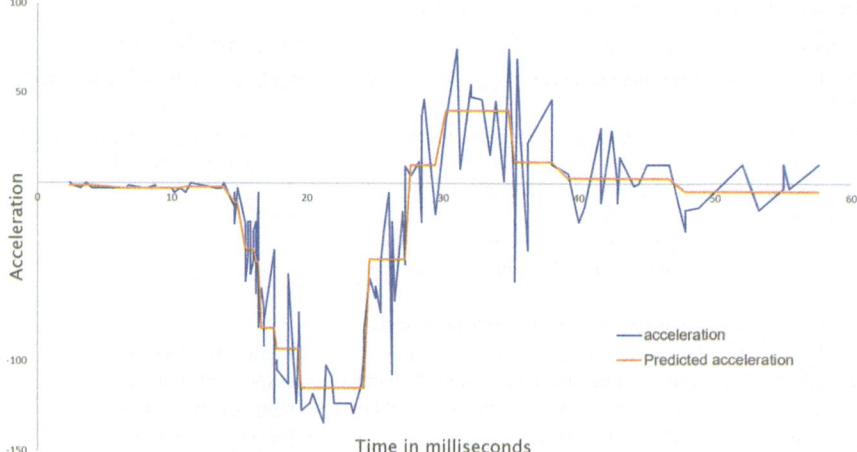

Fig. 8.12 Predicted and actual acceleration vs. time

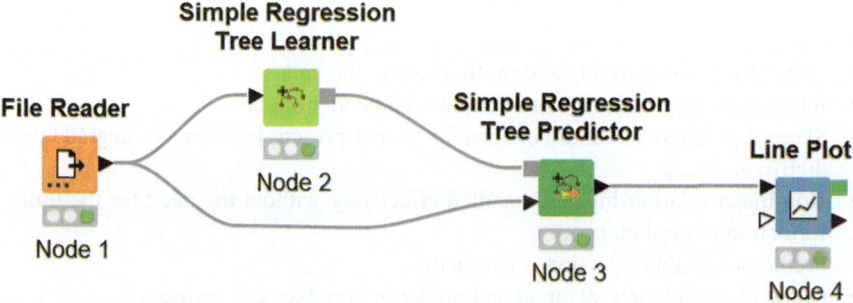

Fig. 8.13 Workflow for regression tree with mcycle.csv data

8.11   Strengths and Weaknesses of Decision Trees

Decision trees are widely used in data mining, but as with every technique, there are both strengths and weaknesses.

Strengths of Decision Trees

- Interpretation is usually straightforward, and the tree can be explained as a series of if-then-else rules.
- The results are displayed in a tree-like structure, which is intuitively appealing.
- Standardization or normalization of predictors is not required.

Table 8.10 Node descriptions for the Fig. 8.13 workflow

Nodes	Descriptions	Input and Output ports
Node 1 File reader	Read file: mcycle.csv	<u>Output port:</u> file table with 133 rows and 2 columns
Node 2 Simple regression tree Learner	This node creates a regression tree Options: Target column: acceleration Include: time Limit number of levels: 10 Check: minimum split size: 15 Check: minimum node size: 5 Check: minimum split node size: 100	<u>Input port:</u> data table from the output port of node 1 <u>Output port:</u> regression tree model
Node 3 Simple regression tree predictor	This node applies regression from a regression tree model by using the mean of the records in the corresponding child nodes to predict acceleration	<u>Input port:</u> model from the output port of node 2 <u>Output port:</u> data table from the output port of node 1 and the predictions from the regression tree model
Node 4 Line plot	Create a line chart of actual and predicted acceleration versus time in milliseconds	Input port: data table from the output port of node 3

- There are few underlying assumptions about the data.
- Interactions among predictor variables can be identified.
- Skewed predictors and outliers usually do not present problems or degrade predictive accuracy.
- Non-linear relationships are handled effectively without the need for the transformation of predictors.
- Predictor variable selection is automatic.
- Categorical and interval target and predictor variables can be used.

Weaknesses of Decision Trees

- Tree algorithms are prone to overfit the data. If the algorithm is allowed to split nodes without constraints, the accuracy of the predictive model will not carry over to new data. So, pruning techniques are important.
- Slight changes in the data set by adding or dropping a predictor or observations can produce dramatically different results.
- Careful validation is required to avoid over-fitting, so large data sets which can be partitioned into separate training and tests are needed.
- Tree algorithms that search exhaustively over all splits tend toward selecting variables that allow more splits, such as continuous predictors and categorical predictors with many levels (Loh & Shih, 1997).
- Considerable judgment and experimentation may be needed to develop suitable results.

8.12 Summary

Decision trees work by recursively splitting the data into more homogeneous groups with respect to the target variable. Two metrics for guiding the selection of splits are available in different tree models, Gini and entropy. The ability of decision tree algorithms to fit complex relationships is both an advantage and a weakness. Therefore, tree growth should be constrained by pre-pruning or post-pruning to avoid overfitting. Common pruning methods include reduced error pruning, cost-complexity pruning, and minimum description length. Error assessment should be made using separate training, testing, and validation subsamples.

Classification trees predict classes based on the dominant class in terminal nodes. Regression trees predict by taking the mean target value in terminal nodes. Applied examples of both types were included and analyzed using KNIME.

References

Aznar, P. (2020). *Decision trees: Gini Vs entropy.* https://quantdare.com/decision-trees-gini-vs-entropy/. Accessed 29 July 2023.

Bramer, M. (2007). *Principles of data mining.* Springer.

Breiman, J., Friedman, L., & Stone, C. (1984). *Classification and regression trees.* Chapman & Hall/CRC.

Esposito, D., & Malerba, F. (1997). A comparative analysis of methods for pruning decision trees. *IEEE Transactions on Pattern Recognition, 19*(5), 476–491.

Galili, T., & I. Meilijson (2020) *Splitting matters: How monotone transformation of predictor variables may improve the predictions of decision tree models.* https://arxiv.org/abs/1611.04561. Accessed 29 July 2023.

Kass, G. V. (1980). An exploratory technique for investigating large quantities of categorical data. *Applied Statistics, 29*(2), 119–127.

Kononenko, I. (1998). The minimum description length based decision tree pruning. In H. Y. Lee & H. Motoda (Eds.), *PRICAI'98: Topics in artificial intelligence* (Lecture notes in computer science. (1531)). Springer. https://doi.org/10.1007/BFb0095272 Accessed 29 July, 2023

Loh, W., & Shih, Y. (1997). Split selection methods for classification trees. *Statistica Sinica, 7,* 815–840.

Pyle, D. (1999). *Data preparation for data mining.* Morgan Kauffman.

Quinlan, J. R. (1987). Simplifying decision trees. *International Journal of Man-Machine Studies, 27*(3), 221–234.

Quinlan, J. R. (1993). *C4.5: Programs for machine learning.* Morgan Kaufmann Publishers.

Quinlan, J. R., & Rivest, R. L. (1989). Inferring decision trees using the minimum description length principle. *Information and Computation, 80*(3), 227–248.

Shannon, C. E. (1948). A mathematical theory of communication. *The Bell System Technical Journal, 27*(3), 379–423.

Sonquist, J. N., & Morgan, J. A. (1964). *The detection of interaction effects.* Survey Research Center, Institute for Social Research, University.

Twala, B. (2009). An empirical comparison of techniques for handling incomplete data using decision trees. *Applied Artificial Intelligence, 23*(5), 373–405.

Wallace, C., & Patrick, J. D. (1993). Coding decision trees. *Machine Learning, 11,* 7–22.

Weisstein, E. W. (n.d.). *Stirling numbers of the second kind.* MathWorld–A Wolfram Web Resource. https://mathworld.wolfram.com/StirlingNumberoftheSecondKind.html. Accessed 29 July 2023

Chapter 9
Naïve Bayes

This chapter presents a predictive model that approximates a Bayesian analysis that can produce quite accurate results. An illustration of how Bayes Theorem can be applied to a simple problem is presented.

9.1 A Thought Problem

Assume that a breathalyzer used by a police department has been shown to incorrectly indicate that a person is over the limit in 5% of the cases. However, the breathalyzer never fails to detect when a person is really over the legal limit. Suppose on a given evening, 1 in 1000 drivers are driving with alcohol over the legal limit. It's New Year's Eve and a traffic checkpoint stop has been set up. Drivers are selected randomly and required to take a breathalyzer test. Assume that a particular driver is found to be over the legal limit for alcohol according to the breathalyzer. Assume that nothing else about the driver is available. What is the probability that the driver is really over the limit? How can the proper likelihood that the person is over the legal limit be correctly estimated? This calls for Bayes' Theorem.

Analysis of the Thought Problem

Bayes' Theorem is a formula describing how to update an event's prior probability when additional evidence is made available. From a Bayesian perspective, prior probabilities are based on the known likelihoods from historical data. In the example of random checks of drivers, the prior is 1/1000 or .001 that the driver is over the legal limit.

© The Author(s), under exclusive license to Springer Nature Switzerland AG 2023   193
F. Acito, *Predictive Analytics with KNIME*,
https://doi.org/10.1007/978-3-031-45630-5_9

The breathalyzer results can update the probability of identifying a driver over the limit. The revised probability is called the posterior probability, which is estimated by Bayes theorem. The goal is to find the probability that the driver is over the limit (OTL) versus (NotOTL), given a positive indication by the breathalyzer indicated. This can be represented as:

$$p\left(OTL \mid POS\right) = \frac{p\left(POS \mid OTL\right) \times p\left(OTL\right)}{p\left(POS\right)} \tag{9.1}$$

where "POS" means that the breathalyzer indicates that the driver is over the limit and is given by $p(POS) = p(POS \mid OTL) \times p(OTL) + p(POS \mid NotOTL) \times p(NotOTL)$.

The following assumptions about the parameters are available from the statement of the problem:

$$
\begin{aligned}
&p\left(OTL\right) = 0.001 \\
&p\left(NotOTL\right) = 1 - p\left(OTL\right) = 0.999 \\
&p\left(POS \mid OTL\right) = 1.00 \\
&\left(the\ breathalyzer\ is\ 100\%\ accurate\ if\ the\ person\ is\ over\ the\ \lim it\right) \\
&p\left(POS \mid NotOTL\right) = 0.05 \\
&\left(the\ breathalyzer\ mistakenly\ reports\ driver\ is\ over\ the\ limit\ 5\%\ of\ the\ time\right)
\end{aligned}
\tag{9.2}
$$

Given the data and a positive indication on the breathalyzer test given to a randomly selected driver, what is the probability that the person is over the legal limit?

The numerator and denominator of Bayes' formula are given by:

$$
\begin{aligned}
&The\ numerator = \\
&\quad p\left(POS \mid OTL\right) \times p\left(OTL\right) = 1.00 \times 0.001 = 0.001 \\
&The\ denominator = \\
&\quad p\left(POS \mid OTL\right) \times p\left(OTL\right) + p\left(POS \mid NotOTL\right) \times p\left(NotOTL\right) = \\
&\quad 1.0 \times 0.001 + 0.05 \times 0.999 = 0.001 + 0.04995 = 0.05095.
\end{aligned}
\tag{9.3}
$$

Substituting the numerator and denominator into Bayes theorem yields:

$$
\begin{aligned}
p\left(OTL \mid POS\right) &= \frac{p\left(POS \mid OTL\right) \times p(OTL)}{p(POS \mid (OTL) \times p(OTL) + p(POS \mid (NotOTL) \times p(NotOTL)} \\
&= 0.001 / 0.05095 = 0.0196
\end{aligned}
\tag{9.4}
$$

When asked for an estimate, many people have answered that the probability that a person is over the limit is as high as 0.95, but the correct probability is about 0.02. This example demonstrates how Bayesian analysis can produce probability estimates that can be quite surprising. The framework of the Bayes theorem can be applied to a supervised analytics problem and be used as a supervised model in data mining to make predictions given a set of predictor variables and a categorical outcome.

9.2 Bayes' Theorem Illustrated

To explain Bayes' theorem pictorially, consider a hypothetical case of a cable service provider with over two million subscribers. The company decided to perform a test market to predict whether current customers will subscribe to a new service. The test involved sending an offer to a random sample of 1000 existing customers. This can be cast as a Bayesian model. Using the test market results, the company would like a predictive model for the rest of its customers.

The test results were that 400 customers bought the new service, so the prior purchase probability was 0.40. This is illustrated in Fig. 9.1, which has a grid representing the 1000 customers in the test.

If this prior probability is applied to the entire subscriber base, the company would expect 400,000 positive responses or 40%. The process of contacting customers via mail, email, and telephone to offer the new service has a cost, so the company wanted to know if there was a way to make the process more efficient than the 40% rate.

Updating the Probabilities

It turned out that the company had data on the gender and age (young or old) of its subscribers, which was also available for the subscribers in the test market. Using gender, the probability of purchase can be refined. There were 600 female customers in the test, 300 of whom subscribed to the new service, and 400 males, 100 of whom subscribed. So, the probability of taking the offer for males is $100/400 = 0.25$, while the corresponding probability for females is $300/600 = 0.50$. The probability was further refined using age, which resulted in the numbers of buyers and non-buyers in each category of age and gender, as shown in Fig. 9.2.

By simply counting the number of customers in each shaded area, the posterior probabilities of each segment could be calculated.

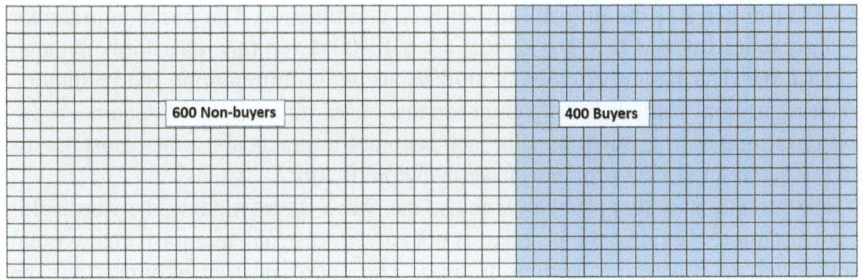

600 Non-buyers　　400 Buyers

Fig. 9.1 Overall results of the test market

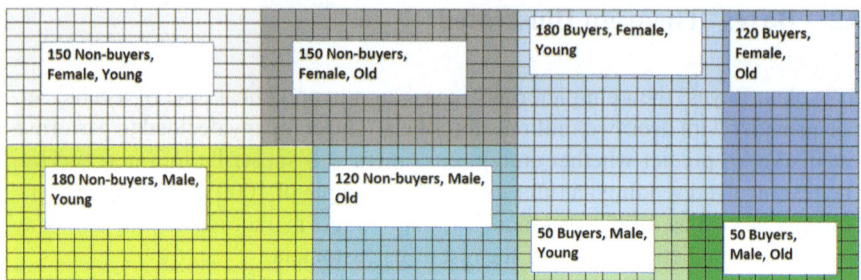

Fig. 9.2 Conditional results of the test market

The probabilities of buying can be easily computed from Fig. 9.2:

- The probability of buying given young and male = 50/(50 + 180) = 0.22.
- The probability of buying given old and male = 50/(50 + 120) = 0.29.
- The probability of buying given young and female = 180/(180 + 150) = 0.54.
- The probability of buying given old and female = 120/(120 + 150) = 0.44.

This small example shows that the Bayes model can be used for predicting the classification of new observations. To classify a new case, find all the observations in the sample with the same descriptive characteristics. With this set of observations, count the number of positive and negative outcomes and apply the counting scheme discussed above. As predictors were added to the analysis, the number of observations used to estimate the proportion taking the offer decreased. For example, the estimate for the proportion of young males subscribing to the service is based on a sample size of 230.

What Happens with More Predictors?

As a more practical example, assume that a predictive problem involves a binary target and 15 binary predictors. This creates $2^{15} = 32,768$ combinations of outcomes and predictor values. Assume that at least 50 observations are needed to make reasonable estimates of the probabilities in each combination. The minimum number of observations would be $50 \times 32,768 = 1,638,400$. Even this may not be enough because the distribution of cases is unlikely to be uniform among the predictors, resulting in combinations having no or too few observations to make accurate probability estimates. The Bayesian approach becomes impracticable since most practical predictive modeling problems have many predictors.

The "solution" to this problem is to use the Naïve Bayes model. The word solution is in quotes because the problem is not solved. Instead, an approximation that works well in practical situations is used. The approximation assumes that the

Table 9.1   The tennis data set

Observation	Play tennis	Outlook	Temperature	Humidity	Wind
1	No	Sunny	Hot	High	Weak
2	No	Sunny	Hot	High	Strong
3	Yes	Overcast	Hot	High	Weak
4	Yes	Rain	Mild	High	Weak
5	Yes	Rain	Cool	Normal	Weak
6	No	Rain	Cool	Normal	Strong
7	Yes	Overcast	Cool	Normal	Strong
8	No	Sunny	Mild	High	Weak
9	Yes	Sunny	Cool	Normal	Weak
10	Yes	Rain	Mild	Normal	Weak
11	Yes	Sunny	Mild	Normal	Strong
12	Yes	Overcast	Mild	High	Strong
13	Yes	Overcast	Hot	Normal	Weak
14	No	Rain	Mild	High	Strong
15	No	Overcast	Hot	High	Strong

predictor variables operate independently of the target. If the predictors operate independently, the joint probabilities of multiple variables can be estimated as the product of the individual probabilities.

9.3   Illustration of Naïve Bayes with a "Toy" Data Set

This example uses a modified version of a small data set consisting of 15 observations with the target variable "play tennis," and weather characteristics thought to affect the decision to play or not play.[1] The observations are shown in Table 9.1.

Calculations Needed for the Naïve Bayes' Model

The probabilities shown in Table 9.2 were obtained by counting, much like was done in with the example of churn above. To get the conditional probability of Sunny given Not playing, note that six observations indicate Not playing. Three of those six observations indicate Sunny, so the conditional probability is 3/6 = .50. Similar calculations were done for each probability in the table. The actual counting was done using an Excel pivot table.

---

[1] https://www.coursera.org/learn/machine-learning-under-the-hood

Table 9.2 Prior and conditional probabilities

Prior probabilities	Play tennis	
	No	Yes
	0.400	0.600
Conditional probabilities	**Play tennis**	
Given outlook	**No**	**Yes**
Overcast	0.167	0.445
Rain	0.333	0.333
Sunny	0.500	0.222
Given temperature	**No**	**Yes**
Cool	0.167	0.333
Hot	0.500	0.222
Mild	0.333	0.445
Given humidity	**No**	**Yes**
High	0.833	0.333
Normal	0.167	0.667
Given wind	**No**	**Yes**
Strong	0.667	0.333
Weak	0.333	0.667

Results of Probability Calculations

To obtain a probability of playing given specific conditions, the naïve Bayes was used. For example, the probability of playing tennis given a Sunny outlook, cool temperature, normal humidity, and strong winds was calculated using the prior and conditional probabilities in Table 9.2. The calculations for one combination of weather conditions are:

$$
\begin{aligned}
&P\left(Yes \mid Sunny, Hot, High, Strong\right) \\
&= \frac{P\left(Sunny \mid Yes\right) \times P\left(Hot \mid Yes\right) \times P\left(High \mid Yes\right) \times P\left(Strong \mid Yes\right) \times P(Yes)}{Numerator + P\left(Sunny \mid No\right) \times P\left(Hot \mid No\right) \times P\left(High \mid No\right) \times P\left(Strong \mid No\right) \times P(No)} \\
&= \frac{0.222 \times 0.222 \times 0.333 \times 0.667 \times .400}{Numerator + 0.500 \times 0.500 \times 0.833 \times 0.333 \times .400} \\
&= \frac{0.006568}{0.006568 + 0.027739} \\
&= 0.192
\end{aligned}
\tag{9.5}
$$

Calculations for all 15 weather conditions were made using the approach of Eq. 9.5. If the probability of "Yes" was greater than. 0.50, the prediction was set to "Yes"; otherwise, the prediction was "No." The results in Table 9.3 show that the naïve Bayes model accurately predicted 14 out of the 15 observations.

Table 9.3 Prediction accuracy using Naïve Bayes

Observation	Play tennis	Outlook	Temperature	Humidity	Wind	Prediction
1	No	Sunny	Hot	High	Weak	No
2	No	Sunny	Hot	High	Strong	No
3	Yes	Overcast	Hot	High	Weak	Yes
4	Yes	Rain	Mild	High	Weak	Yes
5	Yes	Rain	Cool	Normal	Weak	Yes
6	No	Rain	Cool	Normal	Strong	Yes[a]
7	Yes	Overcast	Cool	Normal	Strong	Yes
8	No	Sunny	Mild	High	Weak	No
9	Yes	Sunny	Cool	Normal	Weak	Yes
10	Yes	Rain	Mild	Normal	Weak	Yes
11	Yes	Sunny	Mild	Normal	Strong	Yes
12	Yes	Overcast	Mild	High	Strong	Yes
13	Yes	Overcast	Hot	Normal	Weak	Yes
14	No	Rain	Mild	High	Strong	No
15	No	Overcast	Hot	High	Strong	No

[a]Incorrect prediction

9.4   The Assumption of Conditional Independence

The joint probability of getting a two on a roll of the dice, red on the spinner, and heads on a coin flip can be calculated using the product of the probabilities. Since the three experiments shown in Fig. 9.3 are independent, the likelihood is simply $1/36 \times 1/4 \times 1/2 = 1/48 = .0035$. This is what naïve Bayes analysis assumes about the effects of the individual predictors on the target class in a supervised model, which, of course, is usually not true. However, the assumption is that the predictors are approximately independent.

9.5   Naïve Bayes with Continuous Predictors

For simplicity, the previous examples only had categorical predictors, but Naïve Bayes can be used with continuous predictors. Two approaches can be used for continuous predictors. A simple solution is to discretize the continuous variables into a few categories. However, doing so is sometimes subjective. For instance, in categorizing temperature, someone may select 80 degrees as the cutoff at which temperature can be considered as "High."

In contrast, another person (from the tropics!) may choose to select 90 degrees as the border between "Medium" and "High." This subjectivity causes a noticeable loss of information. But it can still be used as a quick way to get going before applying naive Bayes classification.

Another method is to represent continuous variables with a probability density function. The normal or Gaussian distribution is typically used since it can be

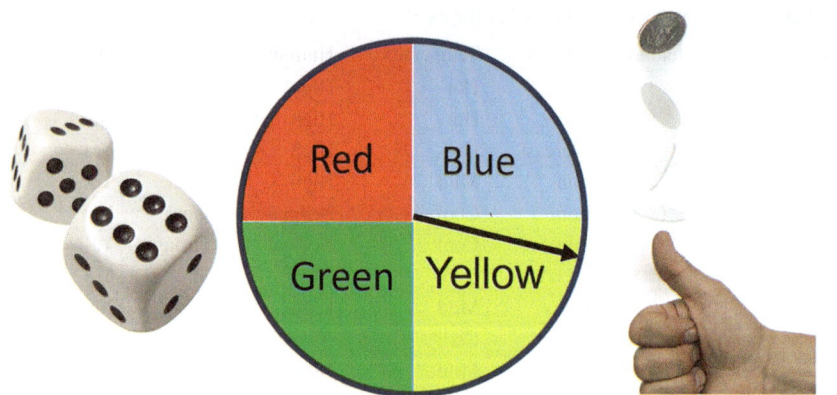

[Marat Musabirov] via iStockphoto.com

Fig. 9.3 Experiments to illustrate independence

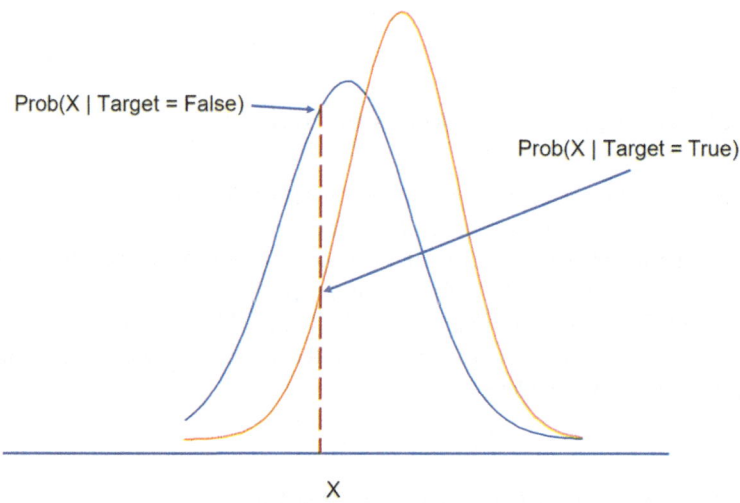

Fig. 9.4 Working with a continuous predictor in naïve Bayes models

represented using the mean and standard deviation. Some software implementations of Naïve Bayes offer the choice of other distribution functions, e.g., Poisson or more complex kernel density functions.

The way this works is demonstrated in Fig. 9.4. Consider a continuous variable, X, which predicts a categorical variable Y which is either True or False. Observations of X are grouped according to the Y values (true or false) in the data sample. The means and standard deviations are computed for each group of X values and used to form the two normal density functions shown in Fig. 9.4. The normal density function only requires a mean and standard deviation. As an example, assume a predictor has a value of X. The conditional probabilities Prob(X|Target = False) and Prob(X|Target = True), which are needed for the naïve Bayes model, are then obtained from the density functions. This method assumes that the normal distribution usefully represents the variable X. At the value of X shown in Fig. 9.4, the probability of False is greater than the probability of True, so X is assigned False.

9.6   Laplace Smoothing

The naïve Bayes algorithm can be problematic in certain situations, especially with small sample sizes. This will happen if a particular value does not occur with a frequency greater than zero in any level of a predictor. When this happens, the conditional probability becomes zero. Since the conditional probabilities are multiplied in a chain, this would cause all posterior probabilities that included the level to be zero. (This was the case in the tennis example illustrated earlier. For the condition of not playing tennis, the overcast level of the weather never occurred.)

A Laplace Smoother (Kuhn & Johnson, 2016) can be used to avoid this. There are several ways to incorporate the smoother, with the simplest being to add one to every count in the combination of predictor values.

9.7   Example of Naïve Bayes Applied to the Heart Disease Data

The data set on predicting heart disease used in earlier chapters was submitted to naïve Bayes. The KNIME workflow is shown in Fig. 9.5, and the node descriptions are in Table 9.4.

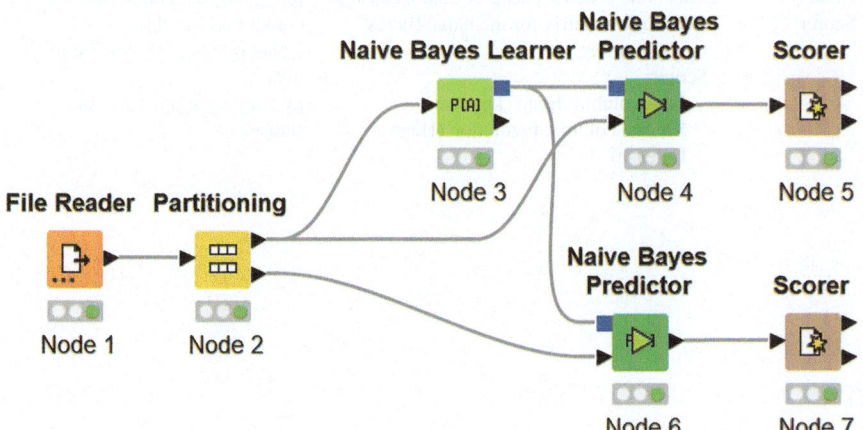

Fig. 9.5 Workflow for naïve Bayes analysis of heart data

Table 9.4 Node descriptions for the Fig. 9.5 workflow

Nodes	Descriptions	Input and output nodes
Node 1 File reader	Read file: ClevelandHeartData.csv	Output port: data table with 297 rows and 14 variables

Node 2
Partitioning
	First partition:
	    Training select relative: 80%
	    Select stratified sampling: Heart_Disease
	    Check: use random seed: 123
		Input port: data table from the output port of node 1
		Upper output port: training data with 237 rows and 14 variables
		Lower output port: test data with 60 rows and 14 variables

Node 3
Naïve Bayes learner
	This node creates a naïve Bayes predictive model
	Options:
	    Classification column: Heart_disease
	    Use defaults for other options
		Input port: data table from the upper output port of node 2
		Upper output port: Naïve Bayes model
		Lower output port: statistics table

Node 4
Naïve Bayes predictor
	Predict the response from the naïve Bayes model for the training set
	Options:
	    Check append individual class probabilities
	    Suffix: prediction (heart disease) training
		Upper input port: model from the upper output port of node 3
		Lower input port: training data
		Output port: classified data with 237 rows and 17 columns

Node 5
Scorer
	This node produces accuracy statistics and a confusion matrix for the naïve Bayes model training set
	Scorer:
	    First column: Heart_Disease
	    Second column: prediction (heart disease)
		Input port: data table from the output port of node 4
		Upper output port: confusion matrix
		Lower output port: accuracy statistics

Node 6
Naïve Bayes predictor
	Predict the response from the naïve Bayes model for the test set
	Options
	    Check append individual class probabilities
	    Suffix: prediction (heart disease) test
		Upper input port: model from the upper output port of node 3
		Lower input port: data table from the lower output port of node 2
		Output port: classified data with 60 rows and 17 columns

Node 5
Scorer
	This node produces accuracy statistics and a confusion matrix for the naïve Bayes model training set
	Scorer:
	    First column: Heart_Disease
	    Second column: prediction (Heart Disease)
		Input port: data table from the output port of node 6
		Upper output port: confusion matrix
		Lower output port: accuracy statistics

Table 9.5 Performance of Bayes model on heart disease data	Partition	Accuracy
	Training	0.852
	Test	0.833

Heart Disease Predictions with Naïve Bayes

The evaluation metrics results for naïve Bayes are shown in Table 9.5.

9.8   Example of Naïve Bayes Applied to Detecting Spam

Email provides a convenient mode of communication used worldwide by millions of people for business and personal messages. However, what started as a curiosity used by techies became the huge number of unsolicited commercial email messages most people receive daily. This soon became, at best, an annoyance and, at worst, a means of deception or even criminal activity.

The proliferation and variety of these unsolicited messages, now called spam or junk mail, led to the development of software programs to detect and screen out such emails. Spam filters have been developed to sift through email messages to separate the "ham" from the "spam." The challenge in designing spam filters is to make the algorithm selective enough to identify spam while not flagging legitimate messages. It has been estimated that about 45% of global email traffic is spam (Spam: Share of Global Email Traffic 2014–2021[2] 2022). Naïve Bayes has been used as the machine learning engine for spam filters because of its simplicity, speed, and accuracy (Karimovich et al., 2020).

A data set of 5556 messages labeled as spam or ham email messages was down-loaded and analyzed using KNIME. Example messages in the data set are in Table 9.6.

The text of each message was prepared by a series of KNIME nodes as indicated in the Text preparation Metanode. The text file was converted into a file consisting of a "bag of words," which creates a separate column for each word in the data set. (The details for converting the text are beyond the scope of this book, but the "bag of words" approach is quite common.) The node descriptions are in Table 9.7, and the KNIME workflow for processing the data is in Fig. 9.6. The workflow created a table with 5572 rows (one for each message) and 12,230 columns with indicators for terms.

---

[2] https://www.statista.com/statistics/420391/spam-email-traffic-share/

Table 9.6   Sample messages in the text data set

Class	Message
Ham	What are you doing? How are you?
Ham	Siva is in hostel aha:-
Spam	Sunshine Quiz! Win a super Sony DVD recorder if you can name the capital of Australia .Text MQUIZ to 82277
Spam	PRIVATE! Your 2003 Account Statement shows 800 un-redeemed S.I.M. points. Call 08718738001 Identifier Code: 49557 Expires 26/11/04

Table 9.7   Node descriptions for the Fig. 9.6 workflow

Nodes	Descriptions	Input and output ports
Node 1 File reader	Read file: spamHam.csv	<u>Output port:</u> data table with 5572 observations and two columns
Node 2 Text preparation	A Metanode containing a sequence of KNIME nodes Nodes in the Metanode are shown below	<u>Input port:</u> data table from the output port of node 1 <u>Output port:</u> 5572 rows by 12,230 columns; the columns were created for each unique word and term found in the data, with the number of occurrences of the word or term in the cell for each message
	Node M2-1: strings to document A document is created for each row	
	Node M2-2: punctuation erasure Remove punctuation from all documents	
	Node M2-3: N chars filter Filter all items in the documents with less than three characters	
	Node M2-4: number filter Remove all digits from the documents	
	Node M2-5: case converter Convert all terms to lowercase	
	Node M2-6: stop word filter Remove all stop words (e.g., "an", "and", "it", "but") from each document	
	Node M2-7: bag of words creator Create a column for each term in the documents. This breaks each message into a set of individual entries with one term or word per entry	

(continued)

Table 9.7 (continued)

Nodes	Descriptions	Input and output ports
	Node M2-8: document vector Create a document vector for each document	
	Node M2-9: category to class Create a new column for each message containing the class (SPAM or HAM)	
	Node M2-10: number to string Convert numbers in columns to strings (for use in naïve Bayes)	
Node 3 Column filter	Column filter to remove the target variable (document) from the input data set Exclude: document Include: all other variables	Input port: data table from the output port of node 2 Output port: table with 5572 rows by 12,229 columns
Node 4 Partitioning	First partition: Training select relative: 70% Select stratified sampling: document class Check: use random seed: 123	Input port: data table from the output port of node 3 Upper output port: training data with 3900 rows by 12,229 columns Lower output port: test data with 1672 rows by 12,229 columns
Node 5 Naïve Bayes learner	This node creates a naïve Bayes predictive model Options: Classification column: document class Use defaults for other options	Input port: data table from the upper output port of node 4 Upper output port: Naïve Bayes model using training data Lower output port: statistics table
Node 6 Naïve Bayes predictor	Predict the response from the naïve Bayes model for the test set Options: Check append individual class probabilities Suffix: prediction (document class)	Upper input port: model from the upper output port of node 5 Lower input port: data table from the lower output port of node 4 Output port: classified test data with 1672 rows by 12,230 columns
Node 7 Scorer	This node produces accuracy statistics and a confusion matrix for the naïve Bayes model test set Scorer: First column: document class Second column: prediction (document class)	Input port: data table from the output port of node 6 Upper output port: confusion matrix Lower output port: accuracy statistics

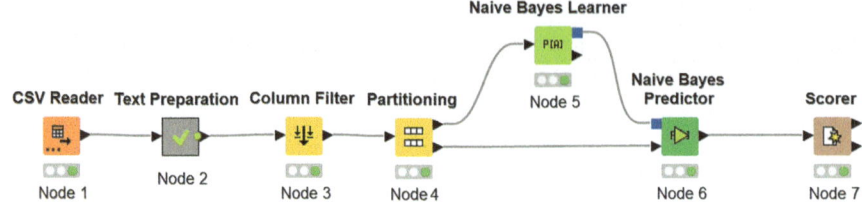

Fig. 9.6 Workflow for SPAM detection using naïve Bayes

Table 9.8 Confusion matrix on test data for naïve Bayes analysis of spam data

	Predicted Y values	
Actual	Ham	Spam
Ham	1461	0
SPAM	17	194
Accuracy = (1461 + 194)/(1461 + 0 + 17 + 194) = .99		

Accuracy of SPAM/HAM Detection

The results of the naïve Bayes analysis of the spam data set in Table 9.8 show good accuracy (99%). As shown in the confusion matrix created for the test data, zero actual ham statements were misclassified as spam, while 17 spam statements were classified as ham. This is a good result since mistakes in filtering out legitimate messages are more serious than mistakenly classifying spam as legitimate.

9.9   Summary and Comments on Naïve Bayes

Naive Bayes is a simple probabilistic classifier that applies Bayes' Theorem assuming predictors are independent. It estimates the probability of each target class given a new observation by using the priors and conditional probabilities derived from the training data. Despite the strong assumption of independence, Naive Bayes often performs surprisingly well. The algorithm has been found to work especially well in applications where there is a large number of predictor variables, possibly because dependence among predictors is less likely.

One weakness is that probability estimates from Naïve Bayes are not always exactly accurate (Rish, 2001). For example, with a binary target of True and False, the estimated probabilities of True and False may be off, but the classification may still be correct.

Compared with a complete Bayesian analysis, Naïve Bayes calculations are practical in situations with smaller sample sizes. Furthermore, naïve Bayes uses all observations in a data set to make estimates, not just those with matching predictor values; this makes the calculations useful in more situations.

Naive Bayes was applied to predicting heart disease detection and identifying spam emails from word frequencies. It achieved 85% and 99% accuracy with test data for the two examples, respectively.

References

Karimovich, G. S., Jaloldin ugli, K. S., & Salimbayevich, O. I. (2020). *Analysis of machine learning methods for filtering spam messages in email services*. International conference on information science and communications technologies, Tashkent, Uzbekistan, pp. 1–4.

Kuhn, M., & Johnson, K. (2016). *Applied predictive modeling* (2nd ed.). Springer.

Rish, I. (2001). *An empirical study of the Naïve Bayes classifier*. IJCAI 2001 workshop on empirical methods in artificial intelligence, pp. 41–46.

Chapter 10
k Nearest Neighbors

K nearest neighbors (kNN) is an interesting, intuitive model used for data mining. Despite the model's simplicity, it can work well with large data sets. It is also one of the top 10 data mining tools in terms of popularity of usage.[1]

KNN is an example of a family of algorithms known as instance-based or memory-based learning that classify new objects by their similarity to previously known objects. KNN does not require many assumptions about the input, predictor data, or error distributions. Formally, it is a non-parametric model. In parametric models (such as multiple regression), assumptions about the error term distribution are needed, and then estimates are made of the distribution parameters. No such considerations are involved with kNN.

A training phase can be used to determine the best value for k, but otherwise, no training is needed for kNN. It can be directly applied to all the data. An important disadvantage of kNN is that the algorithm must be run through all the data available each time a new observation needs to be classified, and this can be time-consuming for large data sets. With logistic regression, naïve Bayes, or OLS, you only save the model and apply it to new data, which is much faster. Since there is no model per se in k-nearest neighbors, all data must be stored and used to analyze new observations.

Typical Applications of kNN
If we look at typical applications, the list is much like other predictive models, such as logistic regression and decision trees, including:

- Flagging fraudulent insurance claims.
- Predicting customer response to promotional offers.
- Selecting the most effective treatment for a medical condition.
- Classifying free-text responses.
- Recommending the next offer to customers in retail settings.

---

[1] https://www.simplilearn.com/10-algorithms-machine-learning-engineers-need-to-know-article

© The Author(s), under exclusive license to Springer Nature Switzerland AG 2023
F. Acito, *Predictive Analytics with KNIME*,
https://doi.org/10.1007/978-3-031-45630-5_10

- Searching for similar documents.

kNN is a method for classifying objects based on similarity. It is called a "lazy" algorithm, which does not use the training data points to generalize and is contrasted with "eager" algorithms. In other words, there is no explicit training phase; if there is one, it is minimal. Most lazy algorithms – especially kNN – make decisions based on the entire data set. Distinctions between lazy and eager learners include:
  With lazy learners:

- The data is stored, not learned from it.
- Classifications are made as soon as a new observation is received.
- The model can be richer since it does not rely on a single pre-specified model.
- The time to classify new observations can be considerable with large data sets.
- Learning time is nonexistent (except for determining the optimal value of k), and minor preprocessing is needed.

With eager learners:

- A model is developed (learned) based on the training data.
- The model is used to classify new observations as they are received.
- The model depends upon a single function derived in the learning phase.
- Classification of new observations is extremely fast, but learning time can be considerable.

10.1   How kNN Works

The kNN algorithm works as follows. A labeled data set with a known categorical target, y (which may be binary or multichotomous), and a matrix X with p potential predictors or features is used. Each case is considered a point in a multidimensional feature space. This allows the computation of "distances" among the points (or cases) based on locations (values) in the feature space.

KNN is then used to classify a new observation described by p variables (x_1, x_2, ..., x_p). The algorithm computes the distance of the new observation to every other observation in the data set. (Euclidean distance is typically used, but other metrics such as the absolute value of differences, squared Euclidean distance, Jaccard distance, Manhattan distance, or cosine distance are also used.) In most cases, it is important to standardize the variables before determining distances to avoid having the variables with the largest numerical values dominating the computed distances.

The distances of the new case to all previous observations are calculated and ranked from closest to farthest. The k existing observations with the smallest distances (the "nearest neighbors") are selected, and a majority "vote" is taken using the category values. The majority vote is used to determine the category of the observation. The number k of nearest points or neighbors to consider in assigning a category is usually odd so that no ties are formed; if an even number is set for k, then random choices for *y* class are made for ties. Since a majority is used to make predictions, kNN can be applied to targets with more than two outcomes.

There is a trade-off implicit here. While only a minimal training phase is used to determine the best *k*, deployment of kNN is inefficient in terms of processing time and memory requirements. More time is needed because all data points are required to classify each new case. This also means that all the data must be stored and available for deployment.

10.2   A Two-Dimensional Graphic Example of kNN

A simple example is used to illustrate how kNN works. Figure 10.1 shows a plot of the characteristics of several types of homes. The two predictors are the area in square feet and the number of rooms, and both variables are standardized to zero mean and unit standard deviation.

Toward the center of the chart is an "unknown" type of home. The two unknown variables, area and the number of rooms, were standardized using the same mean and standard deviation computed for the known data. The distances of the unknown

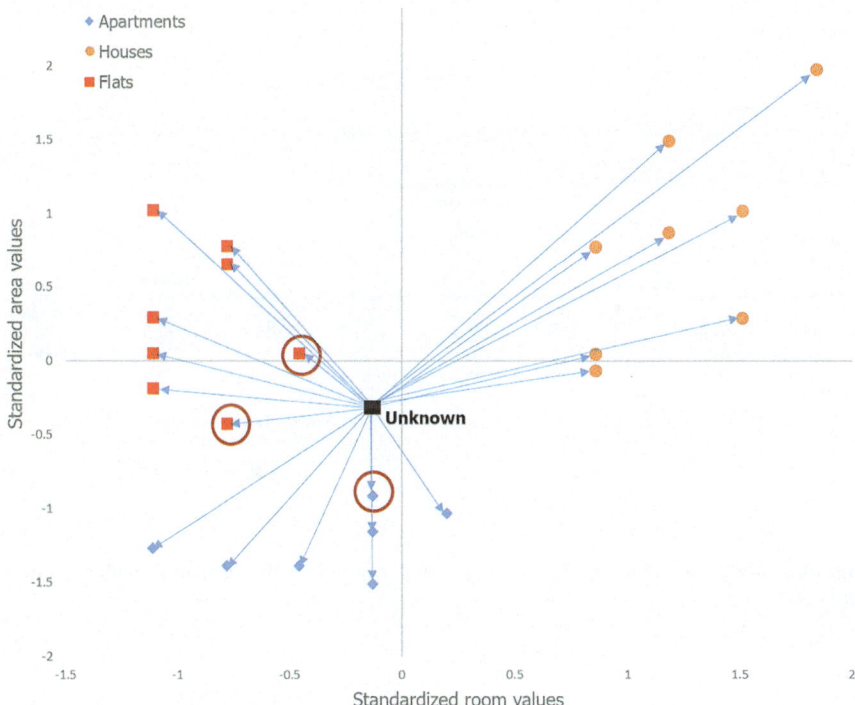

Fig. 10.1 Graphical illustration of kNN

to every data point in the original set are represented by the arrows. The three closest known observations are circled, assuming, for this example, $k = 3$. Two of the closest homes are flats, one is an apartment, and the unknown is classified as a flat using the majority rule.

10.3   Example Application of kNN to Diagnosing Heart Disease

This example uses the database on heart disease from the Cleveland Clinic, which was used in previous chapters. This data set has been used in published studies on machine learning, e.g. (Detrano et al., 1989).

The workflow shown in Fig. 10.2 was first run to determine the value of k that yielded the highest prediction accuracy. Descriptions of each node in the workflow are given in Table 10.1.

Results from the Fig. 10.2 Analysis
The KNIME workflow in Fig. 10.2 determined that the best value for k is 5. A plot of accuracy versus k is shown in Fig. 10.3, which indicates that k = 5 resulted in the highest accuracy, just slightly better than k = 6.

Next, KNN was run on the data with k = 5 resulting in the confusion matrix in Table 10.2.

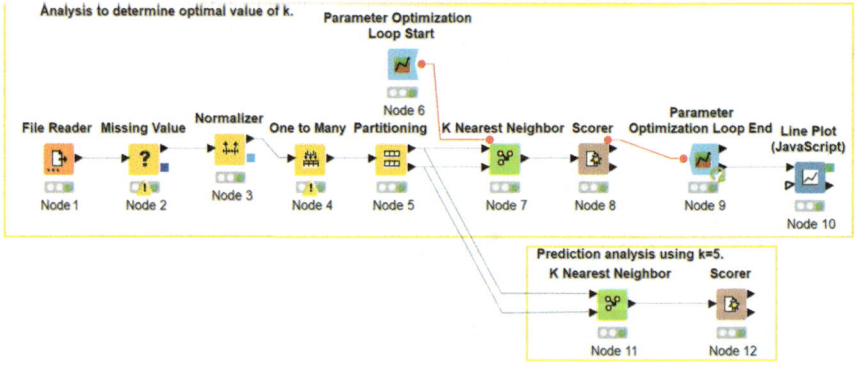

Fig. 10.2 Workflow to find the best value for k and then run prediction analysis with heart disease data

Table 10.1   Node descriptions for the Fig. 10.2 workflow

Nodes	Descriptions	Input and output ports
Node 1 File Reader	Read file: ClevelandHeartData.csv.	Output port: Data table with 297 rows and 14 columns
Node 2 Missing value	Default: Remove Row for all three types of missing values since kNN cannot compute distances with missing values in a row.	Input port: Data table from the Output port of Node 1. Output port: File table with 297 rows and 14 columns (no missing values found).
Node 3 Normalizer	Normalize continuous variables using min-max normalization (0,1) to avoid having high-range variables dominate the distance function in kNN.	Input port: Data table from the Output port of Node 2. Output port: File table with 297 rows and 14 columns with min-max normalization.
Node 4 One to many	KNN only works with continuous variables, so categorical variables are converted to dummy indicators. This node transforms all possible values of categorical variables into separate predictors. The values in each column are either 1, if that row contains the given value or 0 if not. Columns to transform: All string variables except Heart_disease (which is the target variable). (Note: The warning may be ignored in this case.)	Input port: Data table from the Output port of Node 3. Output port: File with min-max normalization and dummy indicators for the categorical variables; 297 rows and 29 columns.

Node 5
Partitioning
	First partition:
	    Select Relative: 80%.
	    Select Stratified sampling: Heart_Disease.
	    Check: Use random seed: 123.
		Input port: Data table from the Output port of Node 4.
		Upper output port: Training data with 237 rows and 29 columns.
		Lower output port: Test data with 60 rows and 29 columns.

Node 6
Parameter Optimization Loop Start
	This node starts an iterative loop to find the best value of k. The red flow variable plot sends the value of k to the k Nearest Neighbors node.
	Standard settings
	    Parameter: k.
	    Start value: 1.
	    Stop value: 12.
	    Step size: 1.0.
	    Check Integer.
		Output: flow variable port with k values

(continued)

Table 10.1   (continued)

Nodes	Descriptions	Input and output ports
Node 7 K Nearest Neighbors	This node classifies a set of test data based on the k Nearest Neighbor algorithm. Standard settings: Column with class labels: Heart_Disease. The k variable input from the optimization loop sets the number of neighbors to consider. Check: Weight neighbors by distance. Check: Output class probabilities.	Input flow variable port: Flow variable from the Output port of Node 6. Upper input port: Data table from the Upper output port of Node 5. Lower input port: Data table from the Lower output port of Node 5. Output port: Test data classified with 60 rows and 29 columns.
Node 8 Scorer	This node produces accuracy statistics and a confusion matrix for the kNN model training set. Scorer: First Column: Heart_Disease. Second Column: Prediction (Heart Disease).	Input port: Data table from the Output port of Node 7. Flow variable port: Accuracy statistics and k value
Node 9 Parameter Optimization Loop End	This node closes an iterative loop to find the best value of k. The red flow variable port sends the value of k to the K Nearest Neighbors node. Options: Flow variable with objective function value: Accuracy. Select: Function should be maximized.	Input flow variable port: Flow variable from the flow port of Node 8. Lower output port: Accuracy values for each value of k.
Node 10 Line Plot	This node produces a line plot of accuracy versus k. Options: Check: Create image at output. Choose column for x-axis. k. Include: Objective value (which will be accuracy). Input port: Accuracy versus k. Axis Configuration: Label for x axis: k value. Label for y axis: Accuracy.	Upper input port: Data table from the Lower output port of Node 9. Lower output port: Line plot of accuracy versus k. (Fig. 10.3)
Prediction analysis using k = 5.		
Node 11 K Nearest Neighbors	This node classifies a set of test data based on the k Nearest Neighbor algorithm. Standard settings: Column with class labels: Heart_Disease. The number of neighbors k set to 5. Check: Weight neighbors by distance. Check: Output class probabilities.	Upper input port: Data table from the Upper output port of Node 5. Lower input port: Data table from the Lower output port of Node 5. Output port: Test data classified with 60 rows and 29 columns.
Node 12 Scorer	This node produces accuracy statistics and a confusion matrix for the kNN model training set. Scorer: First Column: Heart_Disease. Second Column: Prediction (Heart Disease).	Input port: Data table from the Output port of Node 11. Upper output flow port: Accuracy statistics and k value.

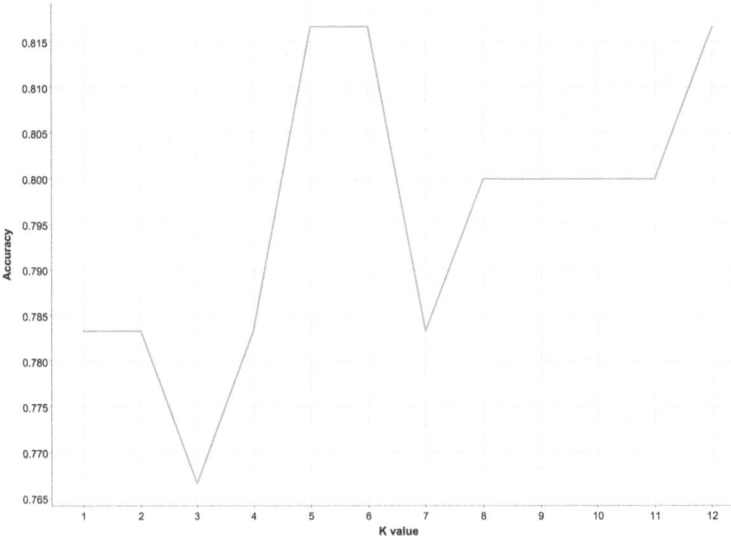

Fig. 10.3 Prediction accuracy vs. k value

Table 10.2 Confusion matrix for predicting heart disease with kNN

	Predicted heart disease	
Actual	not present	present
not present	28	4
present	7	21
Accuracy = (28 + 4)/(28 + 4 + 7 + 21) = 0.817		

10.4   kNN for Continuous Targets

While kNN is primarily a classification method, it can also be used with continuous target variables, much like ordinary least squares (OLS) regression. KNIME does not include a node for kNN regression, so a small R Snippet was created to use the package FNN. One advantage of kNN regression is that non-linear relationships can be easily captured, while ordinary regression requires transformations or adding polynomial predictors to capture non-linearities.

A simple data set with a single predictor (X) and a continuous target(Y) was created. A scatterplot of the data is shown in Fig. 10.4. Note that the relationship is non-linear.

A KNIME workflow was created, shown in Fig. 10.5, to compare kNN and OLS. For this small demonstration data, a test data subset was not used.

The workflow has four sections. The section in the middle left of Fig. 10.5 has nodes to read and normalize the input data. The top section runs a loop to find the value of k, which results in the lowest RMSE. The next section runs kNN with the

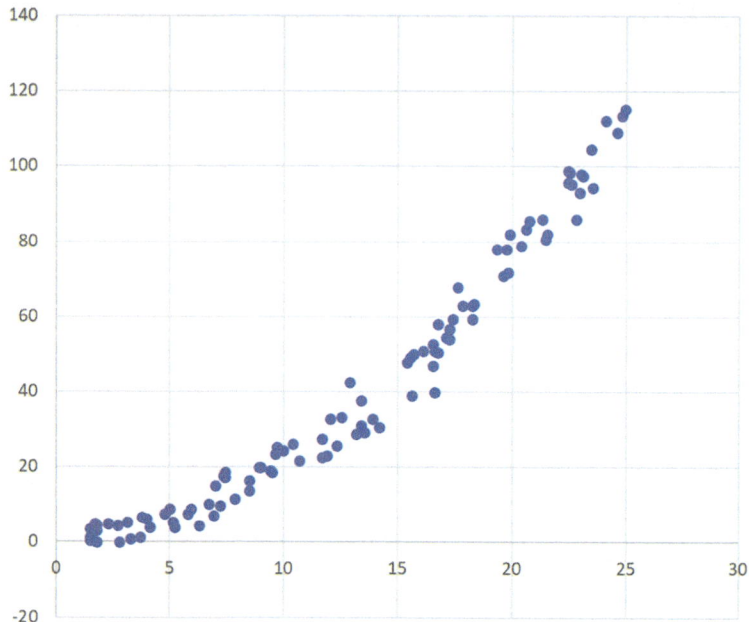

Fig. 10.4 Simulated data with a non-linear X-Y relationship

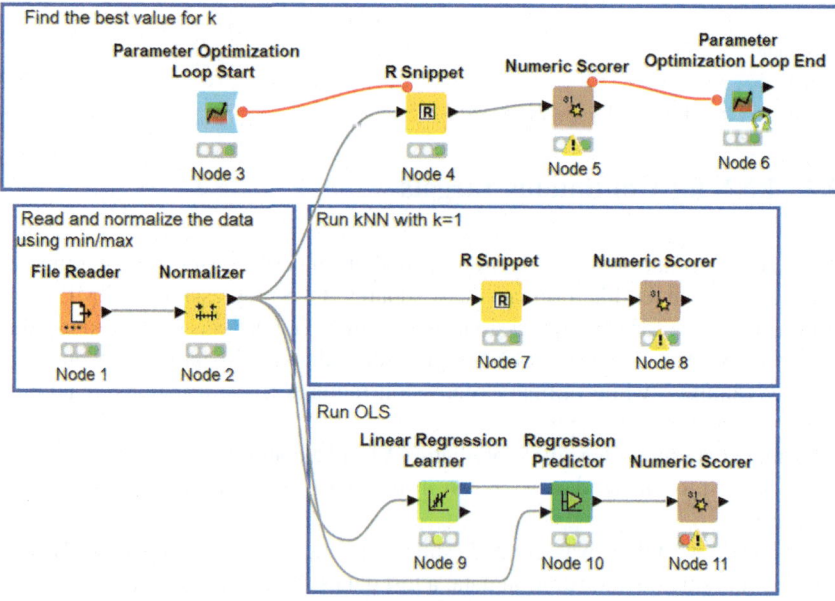

Fig. 10.5 Workflow to compare kNN regression with OLS

optimal value of k, which turned out to be one. The bottom section runs OLS for comparison. Node descriptions for Fig. 10.5 are in Table 10.3.

Results Comparing OLS and kNN with a Non-linear Relationship
Table 10.4 shows the accuracy metrics for the two analyses. OLS does not capture the non-linearity in the data without a transformation, while the kNN model more accurately captures the relationship. The charts in Fig. 10.6 show that the OLS continued to show the non-linearity while kNN (Fig. 10.7) created a linear relationship between the predicted and actual y-values. Of course, OLS could be improved by creating a polynomial model, but the point is that kNN regression did not require a model to be specified a priori.

10.5   kNN for Multiclass Target Variables

KNN is also effective with target variables that have more than two classes. An example data set was obtained from the UCI Machine Learning Repository to illustrate kNN with a multiclass target.[2] The data set has 214 observations, a 6-level categorical target, Type of glass (Table 10.5), and nine continuous predictors (Table 10.6).

The sample size (214) was relatively small, especially considering that the target variable has six levels. So, the distribution of the target values was run for the total sample, the training set, and the test set. The workflow for the distribution analysis is in Fig. 10.8. Node descriptions are in Table 10.7. The partition was 50/50 between the training and test sets, so several observations would be in the less frequent categories for the training and test partitions.

The distributions for the complete data set and the training and test subsets are shown in Table 10.8).

The workflow for the kNN prediction is in Fig. 10.9 and node descriptions for the workflow are in Table 10.9.

The parameter optimization loop reported that $k = 1$ resulted in the most accurate predictions, as shown in Table 10.10. The confusion matrix for predicting type of glass using kNN with the test data set is shown in Table 10.11. The cells with the correct predictions show the frequencies in bold.

The accuracy varies by type of glass, and the overall accuracy is 71.9%. While this may not seem particularly good, predicting a target with six classes with a small data set is a challenging task for a model. For comparison, the model was run five times with the target value shuffled randomly, and the average accuracy with the shuffled data was only 0.2504.

---

[2] German, B., and V. Spiehler. 1987. "Glass Identification Data Set." https://archive.ics.uci.edu/ml/datasets/Glass+Identification

Table 10.3 Node descriptions for the Fig. 10.5 workflow

Nodes	Descriptions	Input and output ports
Node 1 File Reader	Read file: XYNonlinear.csv.	<u>Output port</u>: Data table with 100 rows and two columns.
Node 2 Normalizer	Normalize continuous variables using min-max normalization (0,1) to avoid having high-range variables dominate the distance function in kNN.	<u>Input port</u>: Data table from the Output port of Node 1. <u>Output port</u>: File with 100 rows and two columns with min-max normalization

Node 3
Parameter
Optimization
Loop Start
	This node starts an iterative loop to find the best value of k. The red flow variable plot sends the value of k to the Standard settings:
	    Parameter: k.
	    Start value: 1.
	    Stop value: 10.
	    Step size: 1.0.
	    Check Integer.
		<u>Flow variable output port</u>: Contains an integer value of k.

Node 4
R Snippet
	KNIME does not have a node to perform kNN regression, so R code was inserted in Node 4, which contains an R Snippet. The R code runs kNN for each value of k in the loop.
	The R code is:
	`library(FNN)`
	`mydata <- as.data.frame(knime.in)`
	`Data <- as.data.frame(mydata[,1])`
	`Target <- mydata[,2]`
	`k=knime.flow.in[["k"]]`
	`YKNN - knn.reg(train = Data, test = Data, y = Target, k)`
	`Y_Pred <- YKNN$pred`
	`knime.out <- as.data.frame(cbind(mydata,Y_Pred))`
		<u>Flow variable input port</u>: From the Flow variable output port of Node 3.
		<u>Lower input port</u>: File with 100 rows and two columns and min-max normalization.
		<u>Output port</u>: File table with 100 rows and three columns (X, Y, and Y_Pred).

Node 5
Numeric Scorer
	Score the prediction data.
	Options:
	    Reference column: Y
	    Predicted column: Y_Pred
	Note: MAPE cannot be computed because the min-max normalization creates zeros in the y values. This is not a problem since the criterion for optimization is minimizing RMSE.
		<u>Input port</u>: Data table from the Output port of Node 4.
		<u>Flow variable output port</u>:
		Performance statistics, including RMSE.

(continued)

Table 10.3 (continued)

Nodes	Descriptions	Input and output ports
Node 6 Parameter Optimization Loop End	This node closes an iterative loop to find the best value of k. Options: Flow variable with objective function value: root mean square error. Select: Function should be minimized.	<u>Flow variable input port</u>: Flow variable output from Node 5. <u>Upper output port</u>: Best k value and best RMSE. <u>Lower output port</u>: Accuracy values for all test parameter values in the loop.
Node 7 R Snippet	KNIME does not have a node to perform kNN regression, so R code was inserted in Node 4, which contains an R Snippet. The R code runs kNN with the best value for k, which equals 1. The R code is: `library(FNN)` `mydata <- as.data.frame(knime.in)` `Data <- as.data.frame(mydata[,1])` `Target <- mydata[,2]` `YKNN = knn.reg(train = Data, test = Data, y = Target, k=1)` `Y_Pred <- YKNN$pred` `knime.out <- as.data.frame(cbind(mydata,Y_Pred))`	Input port: Data table from the Output port of Node 2. <u>Output port</u>: File table with 100 rows and three columns (X, Y, and Y_Pred)
Node 8 Numeric Scorer	Score the prediction data from the kNN model with k = 1 Options: Reference column: Y. Predicted column: Y_Pred. Note: MAPE cannot be computed because the min-max normalization creates zeros in the y values. This is not a problem since the criterion for optimization is minimizing RMSE.	<u>Input port</u>: Data table from the Output port of Node 7. <u>Output port</u>: Performance statistics, including RMSE.
Node 9	Linear Regression Learner Settings: Target: Y. Include: X.	<u>Input port</u>: Data table from the Output port of Node 2. <u>Output port</u>: Regression model. <u>Lower output port</u>: Coefficients and statistics.
Node 10	Regression Predictor.	<u>Upper input port</u>: Regression Model from Node 9. <u>Lower input port</u>: Data table from the Output port of Node 2. <u>Output port</u>: OLS predictions plus input data.

(continued)

Table 10.3 (continued)

Nodes	Descriptions	Input and output ports
Node 11 Numeric Scorer	Score the prediction data from the OLS regression model. Options: Reference column: Y Predicted column: Y_Pred Note: MAPE cannot be computed because the min-max normalization creates zeros in the y values. This is not a problem since the criterion for optimization is minimizing RMSE	Input port: Data table from the Output port of Node 10. Output port: Performance statistics, including RMSE.

Table 10.4 Accuracy metrics for OLS vs. kNN

Metric	OLS	kNN
Root mean squared error	0.0754	0.016

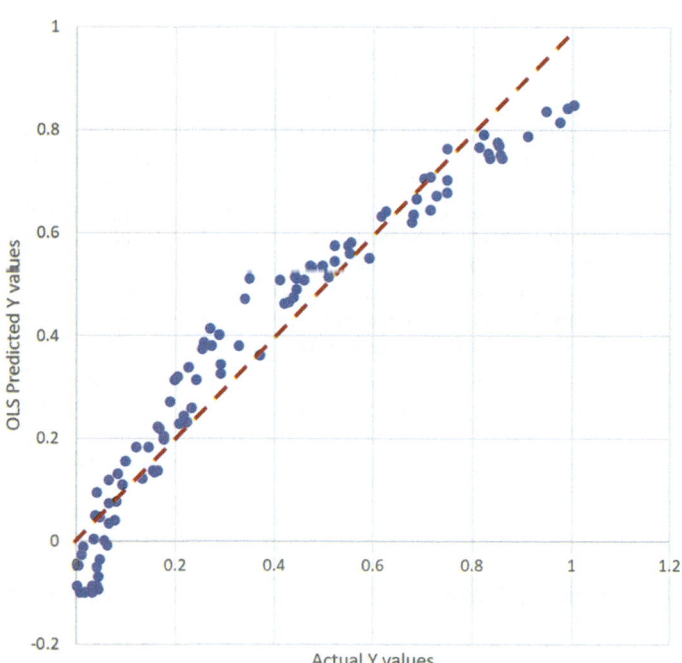

Fig. 10.6 OLS predicted vs. actual values

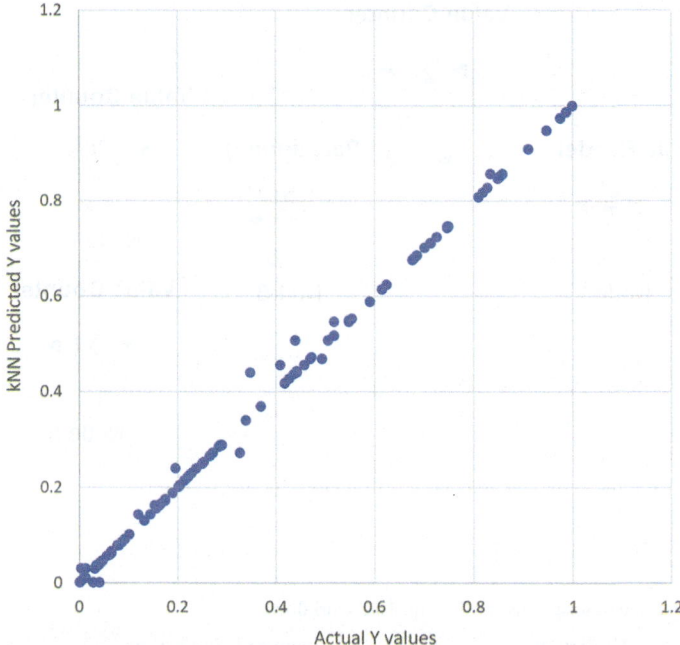

Fig. 10.7 KNN predicted vs. actual values

Table 10.5 Target variable: type of glass

Target variable categories	Description
Building_float	Building windows float processed
Building_nonfloat	Building windows non-float processed
vehicle_windows_float	Vehicle windows float processed
containers	Containers
tableware	Tableware
headlamps	Vehicle headlamps

Table 10.6 Predictor variables in glass identification data

Variable	Description
RI	refractive index
Na	Sodium
Mg	Magnesium
Al	Aluminum
Si	Silicon
K	Potassium
Ca	Calcium
Ba	Barium
Fe	Iron

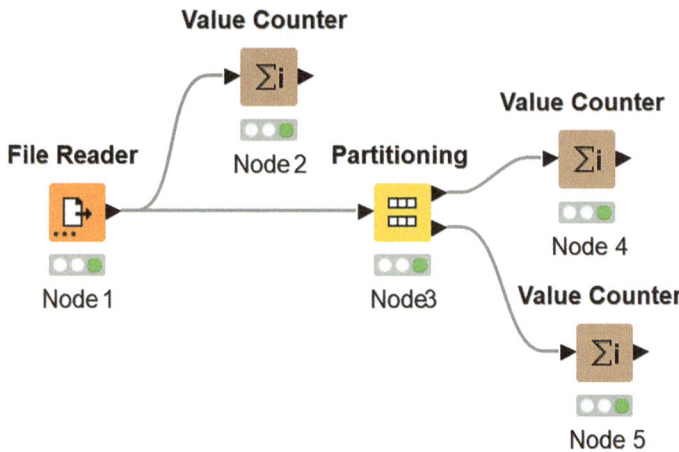

Fig. 10.8 Workflow to find the frequency distributions for glass dataset

Table 10.7 Node descriptions for the Fig. 10.8 workflow

Nodes	Descriptions	Input and output ports
Node 1 File Reader	Read file: glassdata.csv.	Output port: Data table with 214 rows and 10 columns.
Node 2 Value Counter	Settings: Count values for all observations. Column with values to count: Type of glass.	Input port: Data table from the Output port of Node 1. Output port: 6 rows by 2 columns table with frequencies.
Node 3 Partitioning	Settings: First partition. Select Relative: 50%. Select Stratified sampling: Type of glass. Check: Use random seed: 123.	Input port: Data table from the Output port of Node 1. Upper output port: Training data with 107 rows and 10 columns. Lower output port: Training data with 107 rows and 10 columns.
Node 4 Value Counter	Settings: Count values for training set. Column with values to count: Type of glass.	Input port: Data table from Upper output port of Node 3. Output port: 6 rows by 2 columns table with frequencies.
Node 5 Value Counter	Settings: Count values for test set. Column with values to count: Type of glass.	Input port: Data table from Lower output port of Node 3. Output port: 6 rows by 2 columns table with frequencies.

Table 10.8 Frequencies for total, training, and test sets

RowID	Total count	Training set	Test set
Buliding_nonfloat	76	38	38
Buliding_float	70	35	35
headlamps	29	13	16
vehicle_windows_float	17	9	8
containers	13	7	6
tableware	9	5	4

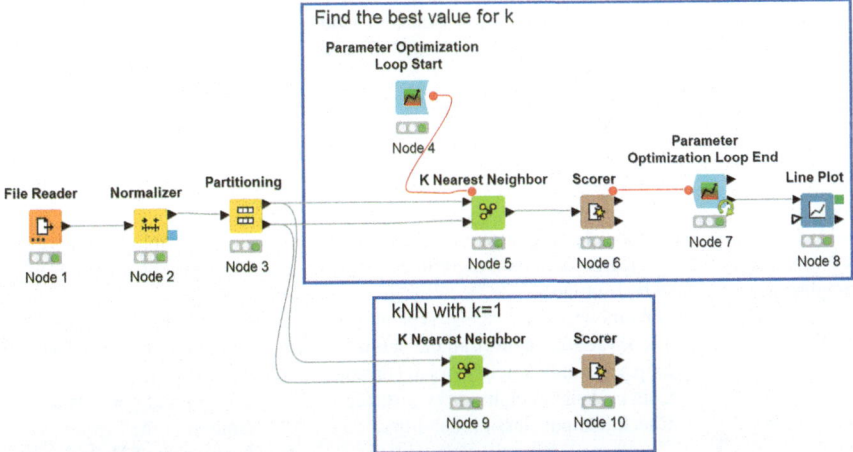

Fig. 10.9 Workflow for kNN analysis of glass dataset

Table 10.9 Node descriptions for Fig. 10.9 workflow

Nodes	Descriptions	Input and output ports
Node 1 File Reader	Read file: glassdata.csv.	Output port: Data table with 214 rows and ten columns.
Node 2 Normalizer	Normalize continuous variables using min-max normalization (0,1) to avoid having high-range variables dominate the distance function in kNN.	Input port: Data table from the Output port of Node 1. Output port: Data table with 214 rows and ten columns with min-max normalization.

(continued)

Table 10.9 (continued)

Nodes	Descriptions	Input and output ports
Node 3 Partitioning	First partition: Select Relative: 60%. Select Stratified sampling: Type of glass. Check: Use random seed: 123.	Input port: Data table from the Output port of Node 2. Upper output port: Training data with 107 rows and ten columns with min-max normalization. Lower output port: Test data with 107 rows and ten columns with min-max normalization.
Node 4 Parameter Optimization Loop Start	This node starts an iterative loop to find the best value of k. The red flow variable plot sends the value of k to the K Nearest Neighbors node. Standard settings: Parameter: k. Start value: 1. Stop value: 12. Step size: 1.0. Check Integer.	Flow variable output port with k values.
Node 5 K Nearest Neighbors	This node classifies a set of test data based on the k Nearest Neighbor algorithm Standard settings: Column with class labels: Type of glass The k variable input from the optimization loop sets number neighbors to consider. Check: Weight neighbors by distance Check: Output class probabilities.	Flow variable input port: Flow variable output from Node 4 Upper input port: Data table from the Upper output port of Node 3. Lower input port: Data table from the Lower output port of Node 3. Output port: Test data classified with 107 rows and 10 columns with min-max normalization.
Node 6 Scorer	This node produces accuracy statistics and a confusion matrix for the kNN model test set. Scorer: First Column: Type of glass. Second Column: Prediction (Type of glass). Show red variable ports.	Input port: Data table from the Output port of Node 5. Flow variable output port: Accuracy statistics. Upper output port: Confusion matrix. Lower output port: Accuracy statistics.
Node 7 Parameter Optimization Loop End	This node closes an iterative loop to find the best value of k. The red flow variable plot sends the value of k to the K Nearest Neighbors node. Connect flow variable port from Node 6 to input flow variable port in Node 7 Options Flow variable with objective function value: Accuracy Select: Function should be maximized	Flow variable input port: Flow variable output port from Node 6. Upper ouput port: Best k value Lower output port: Contains the accuracy for k values in the loop,

(continued)

Table 10.9 (continued)

Nodes	Descriptions	Input and output ports
Node 8 Line Plot	This node produces a line plot of accuracy versus k. Options Check: Create image at output. Choose column for the x-axis: k Include: Objective value (which will be accuracy) Axis Configuration Label for x-axis k value Label for y-axis Accuracy	Input port: Data table from the Lower output port of Node 7. Upper Output port: Chart of accuracy versus *k*.
Node 9 K Nearest Neighbors	Same as Node 5, but with k = 1	Upper input port: Data table from the Upper output port of Node 3. Lower input port: Data table from the Lower output port of Node 3. Output port: Test data classified with 107 rows and ten columns with min-max normalization.
Node 10 Scorer	Same as Node 6	Input port: Data table from the Output port of Node 9. Upper output port: Confusion matrix Lower output port: Accuracy statistics

Table 10.10 Accuracy versus k for kNN prediction of type of glass

k	Accuracy
1	0.720
2	0.682
3	0.645
4	0.654
5	0.626
6	0.664
7	0.645
8	0.645
9	0.636
10	0.654
11	0.617
12	0.617

Table 10.11 Confusion matrix using kNN to predict type of glass

Actual	Predicted X1	X2	X3	X4	X5	X6	Totals	# correct	% correct
X1 = Building float	**24**	6	5	0	0	0	35	24	68.6%
X2 = Building non-float	4	**30**	4	0	0	0	38	30	78.9%
X3 = Vehicle float	1	2	**5**	0	0	0	8	5	62.5%
X4 = Containers	0	2	0	**4**	0	0	6	4	66.7%
X5 = Tableware	0	1	0	1	**2**	0	4	2	50.0%
X6 = Headlamps	2	1	0	1	0	**12**	16	12	75.0%
Totals	31	42	14	6	2	12	107	77	71.9%

10.6   Summary

KNN is a simple, non-parametric classification and regression method that uses similarity or distance measures to make predictions. It finds the k closest training observations to a new observation and assigns the most frequent target value (classification) or average target value (regression).

As demonstrated in the examples, kNN has the flexibility of being able to predict binary, continuous, and multiclass target variables. It also has the advantage of not requiring specific distributions (e.g., normality) for the predictor variables. However, it is necessary to standardize the predictors so that variables with large values do not unduly influence the results.

An advantage of kNN is that it is unnecessary to build a training model and evaluate it with a separate data set. However, a training/test partition is used to select the optimal value of k. While model development is quick, kNN can require long running times if the data set is large. (There are refinements of kNN which can reduce running time (Deotte, 2021; Wang & Wang, 2007).) The long-running time is because the original data and new observations must be used each time predictions are needed. In addition, instead of just storing a model as with regression or other predictive models, the entire data set must be stored and made available for deployment.

Two or more highly correlated variables can result in more weight to the common feature causing the correlations. One common way to avoid this is to submit the predictors to principal components analysis (PCA) and use the component scores on a reduced number of dimensions to compute distances (Clements, 2021). (This also will automatically satisfy the need for standardized values, assuming that PCA is applied to the correlation matrix rather than the covariance matrix.)

Interestingly, the lack of robustness of kNN to noisy data is said to be a weakness by Teknomo (1987), but Ougiaroglou and Evangelidis (2015) claim that kNN is robust to noisy data. Good practice requires that data be carefully prepared and checked for outliers, missing values, and errors before submission to any algorithm.

Finally, kNN suffers from the "curse of dimensionality." The number of dimensions is the number of predictor variables or features, and this refers to the phenomenon of increased distances among observations in high-dimensional spaces. The effect of high dimensionality is that all observations tend to appear equidistant. The resulting sparsity creates problems for kNN and other algorithms because not all combinations of predictor variables are likely to be present in the data. The "cure" for this "curse" is larger data sets. The required number of observations is estimated to grow exponentially with the number of predictors.

References

Clements, J. (2021). *K-Nearest Neighbors (k-NN) explained.* https://towardsdatascience.com/k--nearest-neighbors-k-nn-explained-8959f97a8632. Accessed 29 July 2023.

Deotte, C. (2021). *Accelerating k-nearest neighbors 600X using RAPIDS cuML.* Data Science Technical Blog. https://developer.nvidia.com/blog/accelerating-k-nearest-neighbors-600x-using-rapids-cuml/. Accessed 29 July 2023.

Detrano, R., et al. (1989). International application of a new probability algorithm for the diagnosis of coronary artery disease. *American Journal of Cardiology, 64*(5), 304–310.

Ougiaroglou, S., & Evangelidis, G. (2015). *Dealing with noisy data in the context of k-NN classification.* In Proceedings of the 7th Balkan conference on informatics (pp. 1–4). https://doi.org/10.1145/2801081.2801116. Accessed 29 July 023.

Teknomo, K. (1987). *Strength and weakness of k-nearest neighbor algorithm.* https://people.revoledu.com/kardi/tutorial/KNN/Strength%20and%20Weakness.htm. Accessed 29 July 2023.

Wang, Y., & Wang, Z. -O. (2007). *A fast KNN algorithm for text categorization.* Paper presented at the 2007 International Conference on Machine Learning and Cybernetics. IEEE, Hong Kong.

Chapter 11
Neural Networks

Artificial neural networks are extremely powerful techniques that have become quite popular in recent years. One reason for their popularity is that neural nets are quite flexible and can produce accurate predictions when used in supervised data mining applications. For example, neural nets can be used for problems with categorical target variables in place of or in addition to logistic regression and decision trees. Neural nets also can be used with continuous targets in regression-type problems. Neural nets can evolve into more complex, flexible, and accurate models. The downside is that the models are often difficult to interpret and explain.

Neural nets are especially effective if input variables have non-linear relationships with the target variable. What is fascinating about neural nets is that the model structure needs only to be specified in terms of the number of nodes, the number of hidden layers, and, with certain algorithms, an activation function. The analyst does not have to be concerned with specifying the nature of non-linearities or interactions among predictors.

In a sense, the computer learns from the data using neural nets. A specific model is not specified as with regression models. Instead, the process works like this: "Here's my data; this is how complicated the net can be. Develop a predictive model." These are not statistical models but robust computer programs. Thus, no assumptions are made about normality, linearity, or other properties of the data.

The flexibility and complexity of neural net models are both the source of the attractiveness of neural nets and part of the challenges with effectively using them. Neural nets work best when there are enough observations that large training, validation, and test subsets can be formed.

Neural nets can be easy to apply and use with modern software. The resulting models, when using neural networks, can be quite complicated even though, in one sense, these are just a combination of non-linear regression models. It is the combination of simple models that makes artificial neural nets able to model complicated relationships.

© The Author(s), under exclusive license to Springer Nature Switzerland AG 2023   229
F. Acito, *Predictive Analytics with KNIME*,
https://doi.org/10.1007/978-3-031-45630-5_11

11.1   What Are Artificial Neural Networks?

The "artificial" adjective is used because these models were inspired by attempts to simulate biological neural systems. Data analysts, computer experts, or statisticians did not initially develop the first neural networks. Instead, the original research into human brain activity led to the development of computer models.

The artificial neural net, in a way, mimics human brain neurons. However, the number of elements in even the most complicated neural networks is far from the 100 billion neurons in the human brain. So, artificial neural nets, as used in data mining, are far from as proficient or as complicated as the human brain. Despite this, the terminology persists from the original research on the human brain. The terminology uses terms such as neurons, learning, nodes, activation functions, and synapses in machine-learning neural networks.

Human Neurons
Figure 11.1 is a simplified model of a typical human neuron. In basic terms, the neuron works as follows. The dendrites receive chemical and electrical signals from other neurons. The nucleus processes the information from the dendrites and creates an output which is transmitted by the axon. The axon is then connected via synapses to other neurons. With billions of neurons combined in a network, the result is the powerful capabilities of the human mind.

Mathematical Model of a Neuron
McCulloch and Pitts (1943), neurophysiologists at Yale University, proposed a mathematical model to explain how human neurons work to make decisions and create insights. McCulloch and Pitts (MP) hypothesized that the human brain uses millions of simple elements, essentially on-off switches. Their idea was that a complex set of behaviors, such as those evidenced by humans, can arise from a set of simple units if enough of them act in concert or sequence. The inputs to the MP

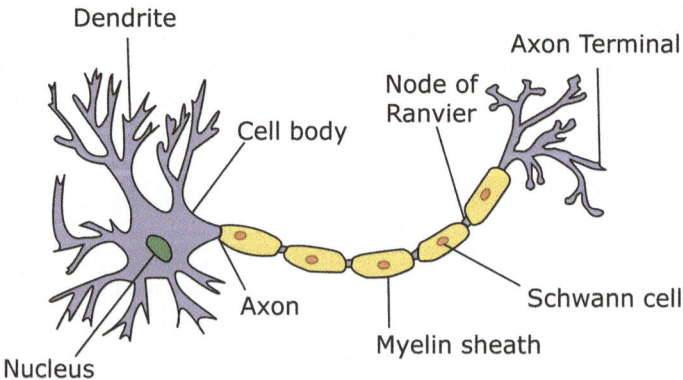

Fig. 11.1 Simplified model of the human neuron. (Image license: User:Dhp1080, CC BY-SA 3.0 <http://creativecommons.org/licenses/by-sa/3.0/>, via Wikimedia Commons)

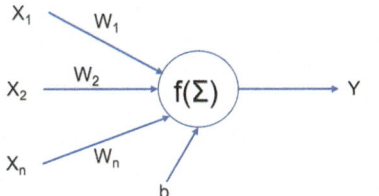

The perceptron model is stated mathematically as:

$$Y = f(\Sigma) = 1 \ if \ S \geq \theta$$
$$= 0 \ if \ S < \theta,$$

$$\Sigma = \sum_{i=1}^{n} W_i X_i + b$$

where $X_1, X_2 \ldots X_n$ are real − valued input variables,
$W_1, W_2, \ldots W_n$ are weights associated with each input,
b is contstant, θ is a threshold, and Y is the output.

Fig. 11.2 An example of a Rosenblatt Perceptron

neuron model were binary, either +1 (excitatory) or −1 (inhibitory). When the sum of the inputs exceeded a threshold, the neuron would "fire."

In 1958 Rosenblatt published an article describing the perceptron, a single-layer neuron, which was built on the MP model. The perceptron allowed inputs to be real numbers (between −1 and +1) and used a weighted sum of the inputs, including a constant bias term (Rosenblatt, 1958). The fundamental elements in the perceptron model are shown in Fig. 11.2.

While Rosenblatt's perceptron had limited success (for example, in image recognition in a device using photocells for inputs), he also contributed to over-hyping the capabilities and potential of neural networks. This hype was the subject of a 1958 article in the *New York Times* that described expectations of "… an electronic computer … that … will be able to walk, talk, see, write, and reproduce itself and be conscious of its existence."[1]

This initial enthusiasm gave way to disillusionment by the late 1960s with the publication of a book by Minsky and Papert (1969) that showed fundamental problems with perceptrons. For example, a single-layer perceptron could not model the so-called XOR (exclusive OR) problem. The XOR case is shown in Table 11.1. Since the single-layer perceptron could not produce the correct output in this table, funding for research into neural nets dried up for more than 10 years, often referred to as the "winter of AI."

It turned out that adding another layer to perceptrons solved XOR problem. Rosenblatt experimented with multilayer models but was unsuccessful in attempts to estimate the weights in such models. A method for "learning" the weights in a multilayer model was needed.[2]

---

[1] "New Navy Device Learns by Doing: Psychologist Shows Embryo of Computer Designed to Read and Grow Wiser." 1958. *The New York Times*, July 8, 1958, pg. 25.

[2] See "A concise history of neural networks"(Jaspreet, 2016) for more on the development of neural nets.

Table 11.1 The
XOR problem

Input A	Input B	Output
0	0	0
0	1	1
1	0	1
1	1	1

11.2   The Learning Process for Artificial Neural Networks

With classical statistical models such as multiple regression, a mathematical model is first developed, and the model parameters are derived analytically by minimizing an error function. In contrast, there is no analytic, closed-form solution to the problem of estimating the weights in a multilayer neural network. Therefore, an iterative process of adjusting the model weights in small increments was needed to minimize an error function. The error function is based on the differences between the observed actual inputs and estimated output target values of the network.

Estimating (training) the weights in a single perceptron can be done straightforwardly, but the estimation problem is more challenging with multilayer networks. One reason is that parameter changes in one layer can affect the best estimates in other layers. In addition, the number of weights can be quite large, and different weights can lead to the same "solution" because of the likelihood of local minima.

Still, multilayer networks had considerable interest because such models could be highly flexible and powerful predictors. After a lengthy period of reduced interest (and available research funding), researchers found ways to estimate weights with these more complex models in the 1980s. Several algorithms were developed which differ in how the weights are adjusted. These developments re-ignited interest in neural nets, and a flurry of research began. One of the first practical methods for estimating weights was the backpropagation algorithm (Rumelhart et al., 1986) and (Sagar, 2019).

Backpropagation requires a network structure, a set of target observations, and input variables associated with each observed value. Understanding the details of backpropagation requires an understanding of matrix calculus, which is beyond the scope of this book. (For an excellent technical exposition, see the online book by Nielsen (2019)). However, in general, the steps in backpropagation are:

1. A starting set of weights is created (randomly) and used to populate the model.
2. The weights propagate through the neural network, creating estimated target values.
3. The differences between the observed and estimated target values are calculated and aggregated into an overall error function.
4. The error is then fed back through the network, and adjustments to the weights that reduce the error are made. The adjustments change the weights in small increments using gradient descent. The gradient measures the direction and magnitude of change in the output function with minor weight changes.
5. After the changes are made, the process returns to step 2 with the adjusted weights.

Steps 2 through 5 are repeated multiple times until a criterion is reached, such as when the changes in the weights are small, a pre-set limit on computer processing time or number of iterations is reached, or until changes in the overall error become negligible.

This process can be slow, sometimes requiring hundreds of iterations, but analyses can often be completed in seconds with modern computers. However, complex problems could take hours or even days. The actual mechanisms are sophisticated, having been developed by mathematicians and computer scientists.

Other methods for training NNs have since been developed, which are sometimes claimed to be superior to backpropagation (Hastie et al., 2009). There are dozens of R packages for neural networks that use various methods (Salsabila et al., 2022). For example, the RSNNS package in R offers standard backpropagation, BackpropMomentum, BackpropWeightDecay, and other methods. KNIME's neural net package uses the RProp algorithm, which has two main advantages over backpropagation: (1) training models is often faster, and (2) there is no requirement for user-specified parameters such as the learning rate. However, Ravichandran (2020) concluded, "Ultimately, none of these … [alternate methods can be considered] … 'better than backprop' because all they do is achieve competitive results." As a result, backpropagation continues to be the most frequently used method.

Perceptron models use an activation function to translate a combination of inputs into an output. The original perceptron model used a step function, whereby a binary output of "1" was produced if the weighted sum of inputs exceeded a threshold, else a "0" was output. Different activation functions have been subsequently used in neural networks. For example, the step function and three other functions are illustrated in Fig. 11.3.[3]

Some neural network software programs use a single activation function while others give the user alternatives. Also, there are implementations that require a single activation function for all layers, while others can have distinct functions in each layer. For a recent review of activation functions, see Mercioni and Stefan (2020).

Early neural net algorithms used the logistic and hyperbolic tangent (TanH) functions, and these are still used in neural net programs with either one or two hidden layers. However, as analysts experimented with networks with three or more hidden layers, a weakness was discovered in using logistic and TanH functions. The ReLU function shown in Fig. 11.3 overcame this weakness enabling networks with three or more layers called "deep learning" models, which are available in KNIME (Silipo & Melcher, 2020).

---

[3] While it is possible to have a linear activation function, most activation functions are non-linear. Using only linear activations would re-create ordinary regression using neural networks.

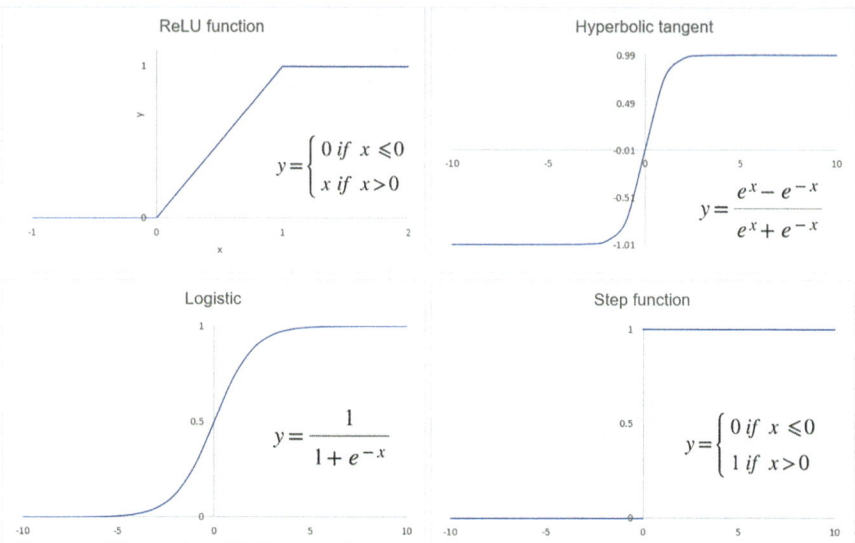

Fig. 11.3 Examples of activation functions used in neural nets

11.3   Example of a Single-Layer Artificial Neuron

The artificial neuron shown in Fig. 11.2 is used to demonstrate the calculations using different activation functions. This model has three inputs, an activation function *f* which acts on a weighted sum of the inputs, and an output Y.

Two different sets of inputs (Case 1 and Case 2, Table 11.2) are used for this demonstration:

For illustration, assume that the weights are as follows in Table 11.3:

The weighted sum for each case is calculated using the following equation:

$$S = b + W_1 X_1 + W_2 X_2 + W_3 X_3 \tag{11.1}$$

Substituting for the weights and the input values, the weighted sums for Case 1 and Case 2, respectively, are:

$$S_1 = -0.1 + 5 \times 1 + 3 \times 3 + 2 \times 2 = 19.0 \tag{11.2}$$

$$S_2 = -0.1 + 5 \times 0.1 + 3 \times -4 + 2 \times 3 = -4.5 \tag{11.3}$$

S1 and S2 are submitted to activation functions ReLU, Logistic, TanH, and Step, and final outputs Y_1 and Y_2 are obtained (Table 11.4):

This demonstration shows how different activation functions affect the outputs of a single perceptron. The activation function will change the specific outputs of neurons, but in practice with one- or two-layer networks, the prediction accuracy will be the same or nearly the same.

Table 11.2 Input values for demonstration

	X_1	X_2	X_3
Case 1	1.0	3	2
Case 2	0.1	−4	3

Table 11.3 Input weights for demonstration

b	W_1	W_2	W_3
−0.1	5	3	2

Table 11.4 Neural net outputs for four different activation functions

	ReLU	Logistic	TanH	Step
Case 1	19.0	1.0	1.0	1.0
Case 2	0.0	0.01	−1.0	0.0

11.4   Example of a Multilayer Perceptron

The real power of neural networks is achieved with multiple nodes connected in layers. The input layer has one or more continuous or nominal variables, and the output layer can have one or more categorical or continuous variables. One or more hidden layers can be created between the input and output layers. Each hidden layer may contain multiple nodes, and the resulting structure is known as a multilayer perceptron.

The diagram in Fig. 11.4 is a multilayer perceptron with an input layer, one hidden layer, and an output layer. A model such as this requires estimating eight weights and three bias levels.

There are three input variables (X_1, X_2, and X_3), and each input is connected to two hidden nodes. The coefficients linking the inputs to the hidden layer are $W_{(1)11}$ through $W_{(1)32}$. The subscript in parentheses indicates the first set of weights, and the next two subscripts indicate the input variable followed by the hidden node. The biases $b_{(1)1}$, $b_{(1)2}$, and $b_{(2)1}$ are labeled in a similar fashion. Outputs $a_{(1)1}$ and $a_{(1)2}$ are computed from the hidden layer after the activation function $f()$ has been applied to the weighted sum of the inputs.

The equations developed for the model in Fig. 11.4 are:

$$a_{(1)1} = f\left(W_{(1)11}X_1 + W_{(1)21}X_2 + W_{(1)31}X_3 + b_{(1)1}\right) \tag{11.4}$$

$$a_{(1)2} = f\left(W_{(1)12}X_1 + W_{(1)22}X_2 + W_{(1)32}X_3 + b_{(1)2}\right) \tag{11.5}$$

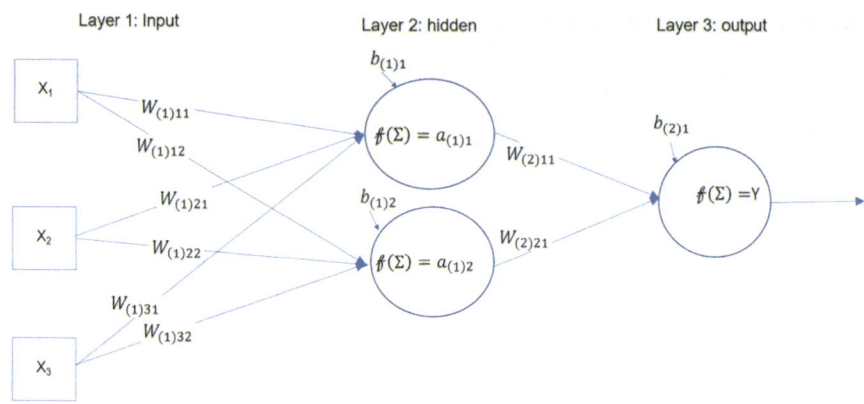

Fig. 11.4 An example of a multilayer perceptron

Table 11.5 Input data

x_1	x_2	x_3	y
0.10	0.02	0.01	0.02
0.20	0.04	0.58	0.04
0.30	0.06	0.70	0.06
0.40	0.08	0.40	0.08
0.50	0.10	0.99	0.10
0.60	0.12	0.10	0.12
0.70	0.14	0.94	0.14
0.80	0.16	1.00	0.16
0.90	0.18	0.30	0.18
1.00	0.20	0.86	0.20

$$Y = f\left(W_{(2)11}a_{(1)1} + W_{(2)21}a_{(1)2} + b_{(2)1}\right) \tag{11.6}$$

The neural net software is tasked with finding values for all the parameters in the model shown in Fig. 11.4. Since there is no closed-form solution to this problem, iterative techniques are used. For the example in Fig. 11.4, assume the data set consists of ten observations (Table 11.5). For this demonstration, the activation function was sigmoid exp.(x)/(1 + exp.(x).

The estimation process begins with the random starting values shown in Table 11.6, along with the final estimated weights.

The objective of the estimation process is to minimize a cost or error function which expresses the difference between the predicted values of the network output and the inputs.

$$C\left(W,b\right) = \frac{1}{n}\sum_{n}^{i=1}\left(y_i - \widehat{y}_i\right)^2 \tag{11.7}$$

Table 11.6 Random starting and final parameter values

Parameter	Random weights	Estimated weights
$W_{(1)11}$	0.013	0.008
$W_{(1)21}$	0.581	4.706
$W_{(1)31}$	0.704	0.060
$W_{(1)12}$	0.400	−0.012
$W_{(1)22}$	0.988	0.582
$W_{(1)32}$	0.100	−0.015
$b_{(1)1}$	0.938	1.811
$b_{(1)2}$	0.997	−0.014
$W_{(2)11}$	0.155	−0.641
$W_{(2)12}$	0.175	0.591
$b_{(2)1}$	0.565	0.704

Table 11.7 Actual and predicted target values for demonstration data

y_i	$\hat{y}_i$	Error
0.02	0.02	0.00
0.04	0.04	0.00
0.06	0.06	0.00
0.08	0.08	0.00
0.10	0.10	0.00
0.12	0.12	0.00
0.14	0.14	0.00
0.16	0.16	0.00
0.18	0.18	0.00
0.20	0.20	0.00

$C(W, b)$ is the cost of predication errors,
W is the set of all the weights in the network,
b is the set of all bias values in the network,
n = the numbr of observations in the training set
y_i = the observations on the target variable,
and, $\hat{y}_i$ = the predicated values of the target variable by the neural network.

For the random starting values, $C(W,b) = 1.928$. After optimization, $C(W,b) = 3.76 \times 10\text{--}5$. The actual values for the target and the estimated values from the model are shown in Table 11.7.

For this simple problem, the errors are 0.0 for all ten observations. In fact, with eleven parameters and only ten observations, the model is overfitting. The process, however, is illustrative of the way neural networks work.

11.5   Example Application of a Multilayer Perceptron with Multi-level Categorical Target

Neural nets can have more than one categorical and continuous output variable. The classic Iris data set is an example with a three-category target. It consists of 150 observations, four measured attributes (sepal length, sepal with, petal length, and petal width), and the type of Iris as the target (setosa, versicolor, and virginica). The four input variables were normalized since the RProp MLP Learner requires numerical inputs from 0.0 to 1.0.

The workflow diagram is shown in Fig. 11.5. The workflow starts by reading the iris data. Then, the data is partitioned into training and test sets. The continuous input variables are normalized to values between 0 and 1, a requirement of the RProp model in KNIME. The predictions from the training and validation data sets are assessed using Scorer nodes. Descriptions of the nodes in the iris workflow are shown in Table 11.8.

Results of the Analysis with the Iris Data
The results of the analysis of the iris data are shown in two confusion matrices, one for the training data (Table 11.9) and one for the test data (Table 11.10). There were no errors with the training data and just two with the test data. Reduced accuracy with the test data is expected since the test data was not used to create the model.

11.6   Considerations for Using Neural Nets

Most neural net programs will drop an observation if missing data is encountered in that observation. The training, validation, and test data must represent the business problem being considered. The old computer science adage "garbage in, garbage out" could not apply more strongly than in neural modeling. If training data is not

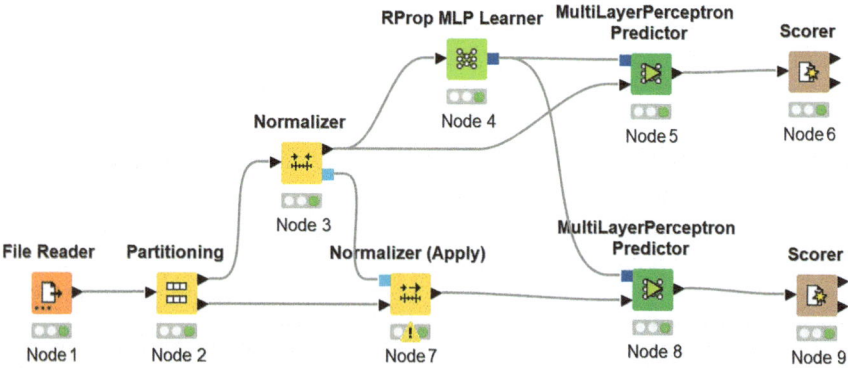

Fig. 11.5 Workflow for neural net analysis of Iris data

Table 11.8   Node descriptions for the Fig. 11.5 workflow

Nodes	Descriptions	Input and Output ports
Node 1 File Reader	Read file: iris.csv.	Output port: Data table with 150 rows and five columns.
Node 2 Partitioning	First partition: 　Select Relative: 70%. 　Select Stratified sampling: 　Species. 　Check: Use random seed: 123.	Input port: Data table from the Output port of Node 1. Upper output port: Training data table with 105 rows and five columns Lower output port: Test data table with 45 rows and five columns.
Node 3 Normalizer	Normalize continuous variables using min-max normalization (0,1) as required by neural nets.	Input port: Data table from the Upper output port of Node 2. Output port: Training data table with 105 rows and five columns with min-max normalization.
Node 4 Rprop MLP Learner	Develop a multilayer feedforward neural network. Options: 　Maximum number of iterations: 100. 　Number of hidden layers: 1. 　Number of hidden neurons per layer: 3. 　Class column: Species. 　Check Use seed for random initialization: Random seed 123.	Input port: Data table from the Output port of Node 3. Model output port: Neural network.
Node 5 MultiLayerPerceptron Predictor	Compute predicted output values from MLP Learner on training data. Options: 　Check Append columns with class probabilities. 　Suffix for probability columns: ProbTraining.	Upper model input port: Model output port from Node 4. Lower input port: Data table from the Output port of Node 3. Output port: Classified data training data table with 105 rows and nine columns with min-max normalization and predictions.
Node 6 Scorer	This node produces accuracy statistics and a confusion matrix for the neural net model with training data. Scorer: 　First Column: Species. 　Second Column: Prediction (Species).	Input port: Data table from the Output port of Node 5. Upper output port: Confusion matrix for training data. Lower output port: Accuracy statistics for training data.
Node 7 Normalizer (Apply)	This node normalizes the input data according to the normalization parameters as given in the model input in the Normalizer Node 3. This normalizes the test data the same way the training data has been normalized in Node 3.	Input port: Data table from the Lower output port of Node 2. Output port: Test data table with 45 rows and five columns with min-max normalization.

(continued)

Table 11.8   (continued)

Nodes	Descriptions	Input and Output ports
Node 8 MultiLayerPerceptron Predictor	Compute predicted output values from MLP Learner on test data. Options: Check Append columns with class probabilities. Suffix for probability columns: ProbTest.	<u>Upper model input port:</u> Model output port from Node 4. <u>Lower input port:</u> Data table from the Output port of Node 7. <u>Output port:</u> Classified data test data table with 105 rows and nine columns with min-max normalization and predictions.
Node 9 Scorer	This node produces accuracy statistics and a confusion matrix for the neural net model with the test set. Scorer: First Column: Species. Second Column: Prediction (Species).	<u>Input port:</u> Data table from the Output port of Node 8. <u>Upper output port:</u> Confusion matrix. <u>Lower output port:</u> Accuracy statistics.

Table 11.9   Neural net results for the iris training data

Training data	setosa	versicolor	virginica
setosa	35	0	0
versicolor	0	35	0
viginica	0	0	35

Table 11.10   Neural net results for the iris test data

Training data	setosa	versicolor	virginica
setosa	15	0	0
versicolor	0	14	1
viginica	0	1	14

representative, then the model's worth is at best, compromised, and at worst, it may be useless. It is worth spelling out problems that can corrupt a training set.

A neural network can only learn from cases that are present. Suppose people with incomes over $100,000 per year might be poor credit risks, and your training data does not include anyone with incomes over $40,000 per year. In that case, you cannot expect the model to make correct decisions on previously unseen cases. Extrapolation beyond the data on hand is dangerous with any model, but neural networks, in particular, may make poor predictions in such circumstances.

A network learns the most accessible features it can. A classic (possibly apocryphal) illustration of this is a vision project designed to recognize military tanks automatically. A network was trained on a hundred images containing military tanks and a hundred without tanks. The model achieved 100% accuracy in predicting which images showed tanks. When assessed on new data, the model proved

Table 11.11   Neural net results for the Iris training data with a randomized target

Training data	setosa	versicolor	virginica
setosa	30	0	0
versicolor	0	31	0
viginica	0	0	29

hopeless. The reason? The pictures of tanks were taken on dark, rainy days, and the pictures without tanks were taken on sunny days. The network learned to distinguish the differences in overall light intensity. Training cases should cover all the weather and lighting conditions under which the model is expected to operate, as well as types of terrain, camera shot angles and distances.

Since a network minimizes an overall error, the proportion of data types in the set is critical. A network trained on a data set with 900 good cases and 100 bad will bias its predictions towards good cases. The network's decisions may be wrong if the representation of good and bad cases differs in a new sample,.

The Overfitting Problem

As with other data mining techniques, the neural net model should be trained on a separate data set and validated and evaluated on different data sets. This is particularly important when using neural nets. Neural nets can predict too well. Given enough flexibility with multiple hidden layers and nodes, a neural net model can be developed to perfectly predict the target in the training data. The problem is that over-fitted models do not generalize well. In other words, when you take a model that fits perfectly to the training data, it may not predict the validation or testing data sets very well. With sufficient iterations and enough nodes, neural nets can even fit data with a randomly created target.

The Iris data set is again used to illustrate overfitting. This time the column with target variables was ordered randomly. This meant that the predictors (sepal length, sepal width, petal length, and petal width) and target (setosa, versicolor, and virginica) were no longer correctly matched. A neural net was specified with two hidden layers, each with ten nodes.

The results are shown in the following two tables. The first confusion matrix is for the randomized training data (Table 11.11). Note that a perfect assignment of the types of Iris flowers was obtained.

The second confusion matrix shows the results using the model on the test data. These results show that the model was overfitting (Table 11.11) and could not accurately predict the new data (Table 11.12).

Overfitting is indicated when the training data results are much better than the validation or test data results. Test and validation results are usually not as good as training results.[4] A frequent question is how much of a reduction in accuracy with

---

[4] It is possible that the accuracy with the test or validation set is higher than with the training set. This should not be a concern since it is most likely due to the randomness in creating the subsets of data.

Table 11.12 Neural net results for the Iris test data with a randomized target

Training data	setosa	versicolor	virginica
setosa	8	7	8
versicolor	2	5	6
viginica	10	7	7

the holdout sample is considered excessive. There are no firm rules or guidelines, and the acceptable difference level depends on the problem. Overfitting by 5% might be unacceptable in specific medical cases, but acceptable if the goal is to predict churn for a direct marketing problem.

If overfitting is judged to be excessive, techniques to reduce it include:

1. Decreasing the model complexity by using fewer nodes per hidden layer or removing a layer.
2. Reducing the number of iterations used for fitting the model.
3. Adding a penalty term to the objective function (i.e., using regularization).
4. Randomly dropping neurons from the network during each iteration of the algorithm.

Techniques (1) and (2) above can be used with the RPropMLP model in KNIME. Techniques (3) and (4) are also available in KNIME, with deep learning models.

Normalizing the Data
Most neural net algorithms require that continuous input and output variables be normalized to between 0.0 and 1.0. In some algorithms, this is done without user input, but the PPropMLP algorithm in KNIME does require normalization. Normalization should be applied equally to the training, validation, and test sets so that predictive accuracy can be accurately assessed. This is done with an applied normalization node in KNIME.[5]

11.7   Example: Using Neural Nets to Predict Credit Status

The widely used German credit data set was corrected in a 2019 report.[6] The report's author also provided some context for the story behind the data.

> The data set contains "… a stratified sample of 1,000 credit ratings (300 bad ones and 700 good ones) from the years 1973 to 1975 from a large regional bank in southern Germany,

---

[5] When applying the normalization from the training set to the validation or test sets, an error is sometimes raised since values found in the latter two data sets might be beyond the range of values in the training set. This is usually not a problem and can be ignored.

[6] Gromping, U. (2019). South German Credit Data: Correcting a Widely Used Data Se. Retrieved from https://archive.ics.uci.edu: https://archive.ics.uci.edu/ml/datasets/South+German+Credit+

Table 11.13 Variables in the German credit data set

Variable	Description
Status	Status of the debtor's checking account with the bank
duration	Credit duration in months
credit_history	History of compliance with previous or concurrent credit contracts
purpose	The purpose for which the credit is needed
amount	Credit amount
savings	Debtor's savings
employment_duration	Duration of debtor's employment with current employer
installment_rate	Credit installments as a percentage of debtor's disposable income
personal_status_sex	Combined information on sex and marital status
other_debtors	Is there another debtor or a guarantor for the credit?
present_residence	Length of time (in years) the debtor lives in the present residence
property	The debtor's most valuable property
age	Age of the debtor in years.
other_installment_plans	Installment plans from providers other than the credit-giving bank.
housing	Type of housing the debtor lives in.
number_credits	Number of credits including (or had) at this bank.
job	Type of debtor's job.
people_liable	Number of people supported by debtor: $1 = 0$ to 2; $2=3$ or more.
telephone	Is there a telephone landline registered in the debtor's name?
foreign_worker	Is the debtor a foreign worker?
credit_risk	Credit score "good" (1) or "bad" (0) in terms of risk.

which had about 500 branches, among them both urban and rural ones. Bad credits are heavily oversampled to acquire sufficient information for discriminating them from good ones; the sources report that the actual prevalence of bad credits is around 5%."

As suggested with the Statlog German credit data,[7] one might consider misclassification cost. It has been recommended for this data that allocating the cost for misclassifying a bad risk as good to be five times as high as the cost for misclassifying a good risk as bad.

The credit data is used to illustrate a neural network application. The data set contains 20 predictors and a binary target: "Score." The Score is either "good" or "bad" based on the debtor's risk assessment before granting credit. The variables in the German credit data set are shown in Table 11.13.

One challenge in developing the analysis is dealing with oversampling. As noted above, the data set is oversampled with 30% "bad" risks while the actual proportion is 5%. This analysis is not designed to maximize predictive accuracy but rather to minimize the expected "cost," which was calculated as five times the number of "bad" risks predicted to be "good" plus one time the number of "good" risks predicted to be "bad."

---

[7] https://archive.ics.uci.edu/dataset/144/statlog+german+credit+data

Overview of the Neural Net Analytic Process

After the data set was read into KNIME, categorical variables were recoded to binary indicators since neural nets must have numeric input variables. Next, the continuous variables were normalized to min/max (0,1) to avoid disproportionate weighting of variables with large values. A partitioning node created two subsets of observations, the first as a training set and the second to assess the model with test data.

A Parameter Optimization Loop was created to tune three neural network parameters: the number of iterations, the number of hidden layers, and the number of nodes per layer. The objective for the optimization was set as "minimization of cost per number of rows in the data table." The reason for dividing by the number of rows in the data table is to make the cost estimates comparable for different-sized tables.

The data table was rescaled to properly assess cost to reflect the correct ratio of 95% "good" and 5% "bad" risks. Costs were set at "1" for incorrectly predicting that a "good" risk is classified as "bad" and "5" for predicting that a "bad" risk is classified as "good."

The expected cost for the lender was estimated using a default assumption that all loan applications would be accepted, that is, not using a predictive model. This is reasonable since 95% of the loans were not in default, and by assigning to the majority, only a 5% error would be made.

The Workflow for the Credit Data Neural Network

The workflow in Fig. 11.6 has three sections; node descriptions are in Table 11.14. The upper section in Fig. 11.6 is the loop set up to parameterize the neural model to minimize cost. The section in the lower right tests the model using holdout data. The section in the lower left calculates the expected cost under the assumption that all predictions are to the majority outcome, which is a "good" risk.

Results of the analysis are shown below in Table 11.15, showing that the cost to the bank is lowered using the model.

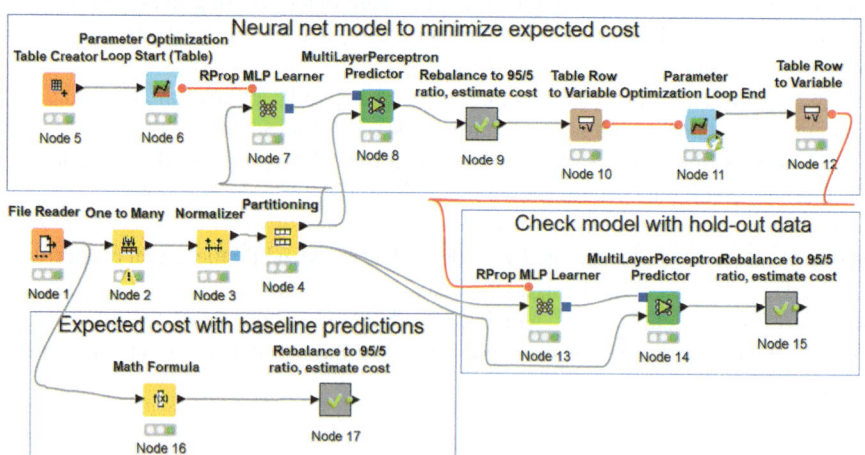

Fig. 11.6 Workflow to tune a neural model to minimize cost

Table 11.14 Node descriptions for the Fig. 11.6 workflow

Nodes	Descriptions	Input and output ports
Node 1 File Reader	Read file:SouthGermanCredit.csv.	<u>Output port:</u> Data table with 1000 rows and 21 columns.

Node 2
One to Many
	This node transforms all possible values in a selected column into a new column. The values in each column are either 1 if that row contains this possible value or 0 if not.
	   Columns to transform: All string variables
	   Check: Remove included columns from the output
	   Do not include original string variables in the output.
		<u>Input port:</u> Data table from the Output port of Node 1.
		<u>Output port:</u> Data table with 1000 rows and 48 columns.

Node 3
Normalizer
	Normalize continuous variables using min-max normalization (0,1) as required by neural nets.
	Settings:
	   Check: Min-Max Normalization
	   Include: All continuous variables
		<u>Input port:</u> Data table from the Output port of Node 2.
		<u>Output port:</u> Data table with 1000 rows and 48 columns with min-max normalization.

Node 4
Partitioning
	First partition:
	   Select Relative: 50%.
	   Select Stratified sampling: credit risk.
	   Check: Use random seed: 123.
		<u>Input port:</u> Data table from the Output port of Node 3.
		<u>Upper output port:</u> Training data table with 500 rows and 48 columns.
		<u>Lower output port:</u> Test and validation data table with 500 rows and 48 columns.

Node 5 Table Creator	This sets the parameter ranges for the neural network cost minimization.	<u>Output port:</u> Manually created data table with three rows and five columns.

	Parameters for loops:
	   No. of iterations:
	      Min: 30.
	      Max: 50.
	      Step: 10.
	   No. of hidden layers:
	      Min: 1,
	      Max: 10.
	      Step: 1.
	   No. of neurons per layer
	      Min: 1.
	      Max: 20.
	      Step: 1.	

(continued)

Table 11.14 (continued)

Nodes	Descriptions	Input and output ports
Node 6 Parameter Optimization Loop Start (Table)	Start a loop to optimize accuracy using the parameter ranges from the output of Node 6. Settings: Parameter name: param_name (string). Start: Min(param). Stop: Max(param). Step size: Step (integer). Parameter value type: type (string)	Input port: Data table from the Output port of Node 5. Flow output port: Parameters.
Node 7 Rprop MLP Learner	Develop a multilayer feedforward neural network. Options: Maximum number of iterations: Controlled by flow variable input. Number of hidden layers: Controlled by flow variable input. Number of hidden neurons per layer: Controlled by flow variable input. Class column: credit risk. Check Use seed for random initialization: Random seed 123.	Flow Input port: Flow output port from Node 6. Input port: Data table from the Upper output port of Node 4. Model output port: Neural network for training data.
Node 8 MultiLayerPerceptron Predictor	Compute predicted output values from MLP Learner on training data.. Options: Check Append columns with class probabilities. Suffix for probability columns: Prob.	Model input port: Model output port from Node 7. Lower input port: Data table from the Upper output port of Node 4. Output port: Classified training data with 500 rows and 51 columns.

(continued)

Table 11.14 (continued)

Nodes	Descriptions	Input and output ports
Node 9 Rebalance to 95/5 ratio, estimate cost	This Metanode rebalances the data to 95% credit_risk = 1 (no default) and 5% credit risk = 0 (default).	Input port: Data table from the Output port of Node 8. Output port: Expected cost per number of rows in the data table.
	Node M9-1 String to Number. B13Change credit_risk and Prediction (credit_risk) from string to number.	
	Node M9-2 Row Splitter. Split credit_risk into two tables – one with credit_risk = 0 (250 rows) and one with credit_risk = 1 (250 rows).	
	Node M9-3 Math Formula. Compute the required number of "bad" risk observations to rebalance to 95/5 split. Expression NBad = COL_SUM($credit_risk$) *.05263157.	
	Node M9-4 Table Row to Variable. Change NBad from table to row variable.	
	Node M9-5 Row Sampling. Draw a random sample of NBad from the credit_risk = 0 table.	
	Node M9-6 Concatenate. Combine the 250 rows with credit risk = 1 with 18 rows with credit risk =0.	
	Node M9-7 Column Expressions. If "credit_risk" = 0 and "Prediction(credit_risk)" = 1, cost =5. If "credit_risk" = 1 and "Prediction(credit_risk)" = 0, cost =1.	
	Node M9-8 Math Formula. Compute the sum of the cost column.	
	Node M9-9 Column Filter. Include Cost and filter all other variables	
	M9-10 Math Formula. Compute cost divided by no. of rows.	
	Node M9-11 Row Filter. Retain only the first row with cost.	

(continued)

Table 11.14 (continued)

Nodes	Descriptions	Input and output ports
Node 10 Table Row to Variable	Change Cost per observation from table to row variable.	Input port: Data table from the Output portof Node 9. Flow variable output: Cost per observation times 100.
Node 11 Parameter Optimization Loop End	Closes the parameter optimization loop and transfers the information back to the corresponding loop start node. Select Function should be: Minimize Cost per observation. Note: Cost per observation scaled by multiplying by 100.	Flow variable input port: Flow variable output from Node 10. Output port: Best parameters.
Node 12 Table Row to Variable	Change best parameters to flow variables.	Input port: Data table from Output port of Node 11. Flow variable output: Best parameters.

Node 13
Rprop MLP Learner
	Develop a multilayer feedforward neural network using the test data and best parameters.
	Options:
	    Maximum number of iterations: Controlled by flow variable input.
	    Number of hidden layers: Controlled by flow variable input.
	    Number of hidden neurons per layer: Controlled by flow variable input.
	    Class column: credit risk.
	    Check Use seed for random initialization: Random seed 123.
		Flow Input port: Flow variable Output port from Node 12.
		Input port: Data table from the Lower output port of Node 4.
		Model output port: Neural network for test data.

Node 14
MultiLayer Perceptron Predictor
	Compute predicted output values from MLP Learner on test data.
	Options:
	    Check Append columns with class probabilities.
	    Suffix for probability columns: Prob.
		Model input port: Model output port from Node 13.
		Lower input port: Data table from the Lower output port of Node 4.
		Output port: Classified training data with 500 rows and 51 columns.

(continued)

Table 11.14 (continued)

Nodes	Descriptions	Input and output ports
Node 15 Rebalance to 95/5 ratio, estimate cost	This Metanode rebalances the data to 95% credit_risk = 1 (no default) and 5% credit_ risk = 0 (default). See details in Node 9.	Input port: Data table from the Output port of Node 14. Output port: Expected cost per number of rows in the data table.
Node 16 Math Formula	Set variable Prediction (credit risk) to 1 for all observations. This predicts all cases to majority value.	Input port: Data table from the Output port of Node 1. Output port: Data table from the Output port of Node 16.
Node 17 Rebalance to 95/5 ratio, estimate cost	This Metanode rebalances the data to 95% credit_risk = 1 (no default) and 5% credit risk = 0 (default). See details in Node 9	Input port: Data table from the Output port of Node 16 Output port: Expected cost per number of rows in the data table X 100.

Table 11.15 Results from neural network analysis

Model	Cost
Cost minimizing predictive model	12
No predictive model	185

11.8   Example: Neural Nets to Predict Used Car Prices

Neural nets can also be used with continuous targets. A subset of the Toyota Corolla data was analyzed using neural nets and, for comparison, ordinary least squares regression. The variables in the Toyota data set are listed in Table 11.16.

The workflow for analyzing the Toyota Corolla data using neural nets and regression is shown in Fig. 11.7. Descriptions of the nodes in the workflow are given in Table 11.17.

Results for Toyota Corolla Analysis with Neural Nets
The results of the analyses show that neural nets performed slightly better than linear regression for this problem in terms of RMSE (Table 11.18). RMSE was computed on the normalized price variable with the test data for both models.

Table 11.16 Variables in the
Toyota data subset

Variable	Description
Price	Price of the used Corolla in Euros
Age	Age of the car (in months)
Mileage	Mileage (in KM)
Fuel Type	Fuel type (Diesel, natural gas, or gasoline)
Horsepower	Horsepower
Automatic	Automatic transmission (Yes or No)
CC	Engine displacement (in CC)
Weight	Weight (in KM)

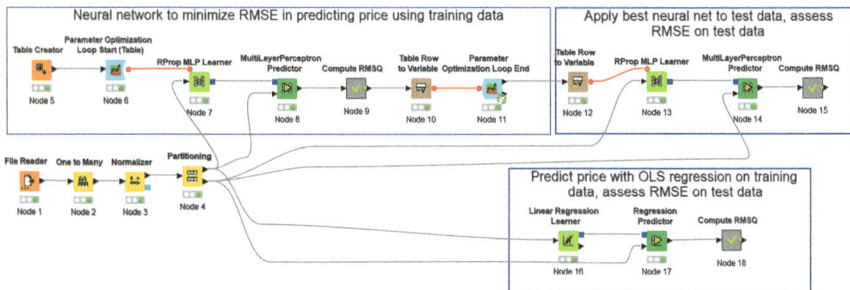

Fig. 11.7 Workflow to predict prices of uses Toyota Corollas

Table 11.17 Node descriptions for the Fig. 11.7 workflow

Nodes	Descriptions	Input and output ports
Node 1 File Reader	Read file: ToyotaCorolla1000.csv.	Output port: Data table with 1000 rows and eight columns.
Node 2 One to Many	This node transforms all possible values in a selected column, each into a new column. The values in each column are either 1 if that row contains this possible value or 0 if not. Columns to transform: Fuel Type. Check: Remove included columns from the output. Do not include the original string variable Fuel Type in the output.	Input port: Data table from the Output port of Node 1. Output port: Data table with 1000 rows and ten columns.
Node 3 Normalizer	Normalize continuous variables using min-max normalization (0,1) as required by neural nets.	Input port: Data table from the Output port of Node 2. Upper output port: Data table with 1000 rows and ten columns with normalized variables. Lower model output port: Normalize model.

(continued)

Table 11.17 (continued)

Nodes	Descriptions	Input and output ports
Node 4 Partitioning	First partition: Select Relative: 70%. Select Draw randomly. Check: Use random seed: 123.	Input port: Data table from the Output port of Node 3. Upper output port: Normalized training data with 700 rows and 10 columns. Lower output port: Normalized test data with 300 rows and 10 columns.

Node 5
Table Creator
	This sets the parameter ranges for the neural network cost minimization.
	Parameters:
	  No. of iterations:
	    Min: 30.
	    Max: 50.
	    Step: 10.
	  No. of hidden layers:
	    Min: 1.
	    Max: 10.
	    Step: 1.
	  No. of neurons per layer
	    Min: 1.
	    Max: 20.
	    Step: 1.
		Output port: Manually created table with three rows and five columns.

Node 6
Parameter Optimization
Loop Start (Table)
	Start a loop to optimize accuracy using the parameter ranges from the output of Node 6.
	Settings:
	  Parameter name: param_name (string).
	  Start: Min(param).
	  Stop: Max(param).
	  Step size: Step (integer).
	  Parameter value type: type (string).
		Input port: Data table from the Output port of Node 5.
		Flow variable output port: Parameters.

Node 7
Rprop MLP Learner
	Develop a multilayer feedforward neural network.
	Options
	  Maximum number of iterations:
	  Controlled by flow variable input.
	  Number of hidden layers: Controlled by flow variable input.
	Number of hidden neurons per layer: Controlled by flow variable input.
	  Class column: Price.
	  Check Use seed for random initialization:
	  Random seed 123.
		Flow variable input port: Parameters from the Flow variable output port of Node 6.
		Input port: Data table from the Upper output port of Node 4.
		Model output port: Neural network for training data.

(continued)

Table 11.17 (continued)

Nodes	Descriptions	Input and output ports

Node 8
MultiLayerPerceptron
Predictor
	Compute predicted output values from
	MLP Learner on training data.
	Options:
	    Check Change prediction column name
	    to Prediction (Price).
		Model input port:
		Model from the Model
		output port of Node 7.
		Lower input port: Data
		table from the Upper
		output port of Node 4.
		Output port: Classified
		training data with 700
		rows and 11 columns.

Node 9 Compute RMSE	Metanode to compute root mean square error between Price and predicted Price.	Input port: Data table from the Output port of Node 8. Output port: RMSE for training data.
	Node M9-1 Math Formula. Compute (Price-Predicted Price) squared.	
	Node M9-2 Math Formula. Compute RMSE+B7.	
	Node M9-3 Column Filter. Retain column RMSE.	
	Node M9-4 Row filter. Retain row 1.	
Node 10 Table Row to Variable	Change RMSE from table to row variable.	Input port: Data table from the Output port of Node 9. Flow variable output port: RMSE for training data.
Node 11 Parameter Optimization Loop End	Closes the parameter optimization loop and transfers the information back to the corresponding loop start node. Select Function should be: Minimize.	Flow variable input port: Flow variable output port from Node 10. Output port: Best parameters.
Node 12 Table Row to Variable	Change RMSE from table to row variable.	Input port: Data table from the Output port of Node 11. Flow variable output port: Best parameters.

Node 13
Rprop MLP Learner
	Develop a multilayer feedforward neural
	network with the test data.
	Options:
	    Maximum number of iterations:
	    Controlled by flow variable input.
	    Number of hidden layers: Controlled by
	    flow variable input.
	    Number of hidden neurons per layer:
	    Controlled by flow variable input.
	    Class column: Price.
	    Check Use seed for random initialization:
	    Random seed 123.
		Flow variable input
		port: Parameters from
		the Flow variable
		output port in Node 12.
		Input port: Data table
		from the Lower output
		port of Node 4.
		Model output port:
		Neural network for
		training data.

(continued)

Table 11.17   (continued)

Nodes	Descriptions	Input and output ports
Node 14 MultiLayerPerceptron Predictor	Compute predicted output values from MLP Learner on test data. Options: Check Change prediction column name to Prediction (Price).	<u>Model Input port:</u> Model output port from Node 13. <u>Lower input port:</u> Data table from the Lower output port of Node 4. <u>Output port:</u> Classified test data with 300 rows and 11 columns.
Node 15 Compute RMSE	Metanode to compute root mean square error between Price and predicted Price. See Node 9 for details.	<u>Input port:</u> Data table from the Output port of Node 14. <u>Output port:</u> RMSE for test data.
Node 16 Linear Regression Learner	Perform linear regression of Price on predictors with training data.	<u>Input port:</u> Data table from the Upper output port of Node 4. <u>Model output port:</u> Regression model predictor.
Node 17 Regression Predictor	Predict Price with test data.	<u>Upper model port:</u> Model from the Model output port of Node 16. <u>Lower input port:</u> Data table from the Lower output port of Node 4. <u>Output port:</u> Classified test data with 300 rows and 11 columns.
Node 18 Compute RMSE	Metanode to compute root mean square error between Price and predicted Price. See Node 9 for details.	<u>Input port:</u> Data table from the Output port of Node 17. <u>Output port:</u> RMSE for test data.

Table 11.18 Accuracy metrics of neural net vs. OLS

Statistics on test data	Neural net	Regression
Root mean squared error	0.042	0.053

11.9 Summary

Artificial neural networks (ANNs) are predictive models inspired by biological neural networks. They can model complex non-linear relationships between inputs and outputs. ANNs learn from training data by iteratively adjusting weights and bias levels to minimize an error function. The ANN structure contains an input layer, one or more hidden layers, and an output layer. Different activation functions like logistic, tanh, ReLU can be used to introduce non-linearity.

ANNs can handle non-linear problems and interactions without explicitly specifying the model structure. No distribution assumptions are needed, and they can be used for categorical or continuous criterion and predictor variables. Training ANNs can be computationally intensive, and the resulting model, often quite complex, is treated as a "black box," making interpretation of the model difficult. No p-values to assess specific predictors or overall model performance are available.

Overfitting is a risk, requiring training/validation/testing data splits. Model complexity and training iterations need controlling. Data pre-processing, like normalization, is often needed and missing values cannot be handled directly.

References

Hastie, T., Tibshirani, R., & Friedman, J. (2009). *The elements of statistical learning*. New York: Springer.

Jaspreet (2016). *A concise history of neural networks*. Retrieved from towardsdatascience.com: https://towardsdatascience.com/a-concise-history-of-neural-networks-2070655d3fec

McCulloch, W. S., & Pitts, W. H. (1943). A logical calculus of the ideas immanent in nervous activity. *Bulletin of Mathematical Biophysics, 5*, 114–133.

Mercioni, M. A., & Stefan, H. (2020) *The most used activation functions: Classic versus current*. In 2020 International conference on Development and Application Systems (DAS), Suceava, Romania, pp. 141–145.

Minsky, M., & Papert, S. (1969). *Perceptrons: an introduction to computational geometry*. MIT Press.

Nielsen, M. (2019). *Neural networks and deep learning*. http://neuralnetworksanddeeplearning.com/. Accessed 29 July 2023.

Ravichandran, Z. (2020). *Backpropagation and its alternatives*. https://medium.com/@sallyrobotics.blog/backpropagation-and-its-alternatives-c09d306aae4c. Accessed 29 July 2023.

Rosenblatt, F. (1958). The perceptron: A probabilistic model for information storage and organization in the brain. *Psychological Review, 65*(6), 386–408.

Rumelhart, D. E., Hinton, G. E., & Williams, R. J. (1986). Learning internal representations by error propagation. In D. E. Rumelhart (Ed.), *Parallel distributed processing: explorations in the microstructures of cognition* (pp. 318–362). MIT Press.

Sagar, V. (2019). *5 techniques to prevent overfitting in neural networks*. Retrieved from www.kdnuggets.com, https://www.kdnuggets.com/2019/12/5-techniques-prevent-overfitting-neural-networks.html. Accessed 29 July 2023.

Salsabila, M., A. Verma, C. Dutang, P. Kiener, and J. Nash. (2022). *A review of r neural network packages (with NNbenchmark): Accuracy and ease of use*. https://www.inmodelia.com/exemples/2021-0103-RJournal-SM-AV-CD-PK-JN.pdf. Accessed 29 July 2023.

Silipo, R., & Melcher, K. (2020). *Codeless deep learning with KNIME*. Packt Publishing.

Chapter 12
Ensemble Models

Many different machine-learning models have been developed, and new variants continue to be explored. This leads to questions such as, "Which algorithm is best?" and "Which algorithm is best for a particular problem context or data set?" As expected, there is no simple, definitive answer to these questions. An interesting comparison of the performance of five algorithms using six different data sets found that none performed best on all six data sets (Seni & Elder, 2010).

Observations of model performance led to combining the predictions from two or more models, a process called ensemble learning. Ensemble models can lead to greater accuracy when the predictions of diverse individual models are independent and aggregated to make a final estimate. The effectiveness of combining independent estimates was the theme of the book *The Wisdom of Crowds* (Surowiecki, 2005).

Wallis (2014) described a famous example of the power of combining estimates. In 1907 Sir Francis Galton published an article in *Nature* about combining individual estimates. Galton attended the West of England Fat Stock and Poultry Exhibition, where he was intrigued by a weight guessing contest. The goal was to guess the weight of an ox when it was butchered and dressed. Around 800 people entered the contest and wrote their guesses on tickets. The person who guessed closest to the butchered ox's weight won a prize.

After the contest, Galton borrowed the tickets and discovered that the median guess across all entrants was 1207 pounds, only 9 pounds from the actual weight of 1198 pounds. (The mean value of the guesses was exactly 1198.)

© The Author(s), under exclusive license to Springer Nature Switzerland AG 2023
F. Acito, *Predictive Analytics with KNIME*,
https://doi.org/10.1007/978-3-031-45630-5_12

12.1   Creating Ensemble Models

To create ensemble models, two or more learning models that attempt to predict the same outcome are needed. Creating diverse learners that can be combined to create an ensemble model can be done by:

- Using two or more algorithms, such as decision trees, logistic regression, and neural networks.
- Changing the parameters of a single model, such as the number of branches in a decision tree or the number of nodes in neural networks.
- Sampling different subsets of predictor variables to create multiple models.
- Sampling observations to create multiple models.

Benefits of Ensemble Models
Ensemble models have two main benefits: improved prediction accuracy and reduced variation (reliability) in predictions with different data samples and (Brownlee, 2020). Whether ensembles are created by running a single model on multiple data samples or running multiple models on a single data set, different predictions are likely. If the errors across samples or models are independent, or at least nearly independent, combining predictions will tend to cancel errors and lead to greater accuracy.

While assessments of prediction accuracy are important, it is also essential to consider the variation as different data samples are analyzed. One way to assess this variation is to use a k-fold partitioning of a data set. The performance will typically vary if k samples are taken and submitted to the tree model. The range or standard deviation of this variation indicates performance expected with unseen data. An ensemble model may not improve the average performance across k samples compared with a single model and yet have a smaller variation in performance.

12.2   Ensemble Models Based on Decision Trees

Decision tree models have been available for years, but their accuracy and usefulness have increased with the development of ensemble models. Decision trees evidence instability if the data set is changed slightly. This behavior is due to two characteristics of tree models. (a) The models produced are not globally optimum. Once a split is done, no further consideration is given to altering that split based on divisions being made further down the tree. (b) If two potential predictor variables, A and B, are highly correlated, then if variable A is selected for a split, this effectively makes variable B unimportant. That is to be expected. However, if a slightly different sample is selected, for example, by randomly splitting a data set into training and test subsets, variable B might be chosen first, making variable A less important. Thus, changing the seed for the random split could result in different results.

Tree ensemble models require two steps:

- Creating a set of distinct predictive tree models.
- Combining the predictions from the collection of models to produce an overall prediction.

Two approaches to creating tree-based ensemble models differ in how the multiple models are created. Parallel learners generate a set of strong (or complex) models, and the independence of the models is used to average out errors. Examples include Bagging and Random Forests. With sequential learners, consecutive weak models are created, and errors are captured at each stage of the process. The errors at each stage become the target variables for the next step. Examples of sequential ensemble models include AdaBoost, Gradient Tree Boosting, and XGBoost.

Bagging

Bagging multiple data sets by bootstrap sampling and then aggregating the results to create predictions is an early parallel ensemble approach (Breiman, 1966). With a bootstrap sampling of n observations, k repeated samples of size n are selected with replacement. Figure 12.1 shows the bagging approach. Repeated sampling means that observations are left out of a given sample, and some observations are included more than once. For regression trees, the average of k predictions across becomes the ensemble prediction. A majority vote is taken for classification trees to obtain the predicted class.

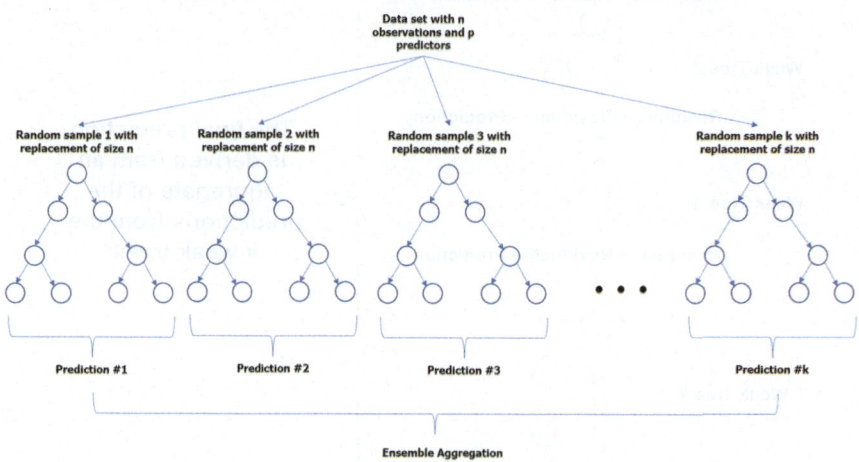

Fig. 12.1 Diagram of a bagged ensemble tree model

Random forests, another example of parallel ensemble models, introduce another element of sampling to increase the diversity of models created (Ho, 1998). Instead of using all available predictors to make each tree, a sample of predictor variables is taken. So, two forms of randomization are involved: sampling observations with replacement and sampling of features. The diagram for random forests would appear like that for bagging shown in Fig. 12.1, except that each tree would be created with a randomized subset of the predictors.

Boosting

Boosting is building a large, iterative decision tree by fitting a sequence of smaller decision trees called layers. The tree at each layer consists of a small number of splits. Each tree can be tiny, with only a few partitions. The process for boosting is diagrammed in Fig. 12.2.

An early boosting algorithm, called AdaBoost, was developed by Freund and Schapire (1996) and used for categorical target variables. However, subsequent extensions to continuous targets were implemented.

Two aspects of the AdaBoost model make this a powerful method. First, a weak model is created at each stage, and the prediction errors are captured. The errors from each step become the target for the next stage. Thus, the overall prediction accuracy improves as the model faces increasingly difficult predictions. The second

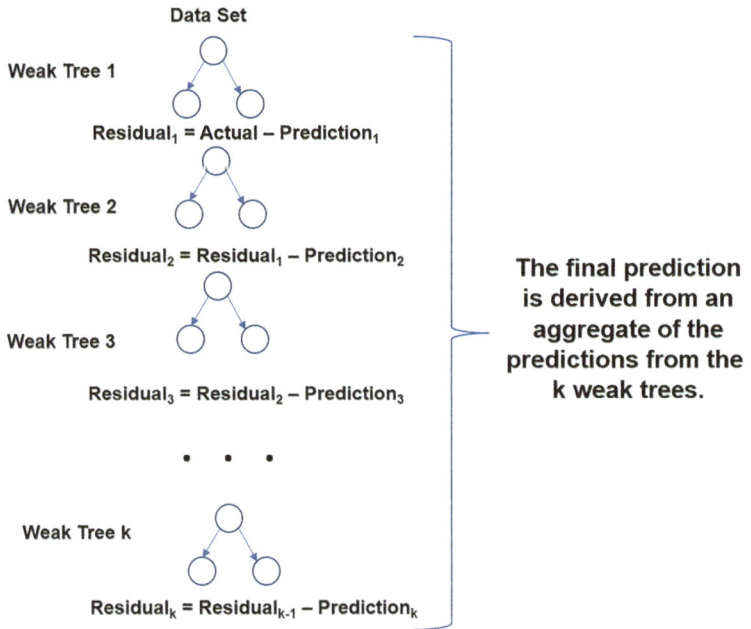

Fig. 12.2 Diagram of a boosted tree ensemble model

important aspect of AdaBoost is that at each stage, the observations that result in prediction errors are given more weight compared with observations that are correctly predicted. In the initial stage, all weights are equal. At the end of the process, AdaBoost creates a weighted ensemble of the predictions made at each, with weights proportional to the accuracy of each stage.

Gradient Boosting Machines (GBM) build on the approach of the AdaBoost algorithm by incorporating gradient descent for optimization (Friedman, 2001). These models have been popular and successfully applied in multiple contexts such as time-to-event projections, Poisson regression, and multinomial classification.

Weights are calculated for each predictor variable that will minimize prediction error, called a "loss function" in the context of the GBM models. A typical loss function for continuous variables is based on the square of actual minus predicted values. For binary responses, the loss function is the negative of the log-likelihood. A brute force approach would be to form a grid of incremental values of each predictor, but such an unguided search process would be grossly inefficient in building a tree model. Instead, a calculus-based approach to finding a local minimum is based on the error between the actual and predicted values of the target variable. The gradient (sort of multivariable slopes or derivatives) directs changes in the model parameters with the steepest slopes toward minimization.

The minimization process proceeds iteratively in small steps. (The user can set the step size. Large steps will reduce computer processing time but may miss a local minimum. Small steps are more likely to find a local minimum at the cost of increased time.) The number of iterations required depends on the problem and the step size and can range from a few to thousands or more. This method does not guarantee a global minimum, so algorithms typically incorporate multiple runs through the data with different starting values.

An updated model (Friedman, 2002) inserted a step before constructing each tree in the sequence, which took a sample of the data. The tree was formed on this sample, and the loss function was calculated. At the next iteration, another random sample of the entire data set is selected, and so on. This increased the accuracy of the predictions as compared with GBM.

XGBoost (which stands for "eXtreme Gradient Boosting") (Chen & Guestrin, 1996) has become one of the most popular ensemble modeling techniques in recent years, primarily due to its use in Kaggle competitions (Adebayo, 2020). It is considered "state of the art" by some practitioners. XGBoost is very flexible and customizable, and it is open source, runs on many operating systems, provides options such as regularization to reduce overfitting, and includes many parameters to tune the algorithm to specific applications. Furthermore, it executes faster than most comparable models.

12.3   Example of Ensemble Modeling with a Continuous Target

The file ToyotaCorolla.csv[1] contains 1436 records of used cars on sale during the summer of 2004 in The Netherlands. In addition to the price of each automobile, there are details on more than 30 attributes, such as the weight of each car, age, number of kilometers, horsepower, and optional accessories such as power steering, central locking, and tow bars.

A KNIME workflow (Fig. 12.3) was used to create and compare ordinary least regression (OLS) and Gradient Boosted Trees to predict the prices of the Toyota Corollas. Training and test data sets were made, and performance metrics were saved. The descriptions of each node are shown in Table 12.1.

The results in Table 12.2 show that the Gradient Boosted Trees performed better than ordinary regression on all three performance measures.

12.4   Example of Ensemble Modeling for a Binary Target

This example uses the data set on churn used in Chap. 8. This time the XGBoost Ensemble Learner model is used. The goal is to find the best value of the threshold variable that minimizes the expected cost to the telecom company using the following cost assumptions:

Table 12.3 indicates that predicting "no churn" when the actual case is "churn" is 100 times the cost to the company. The workflow for this example is in Fig. 12.4 and the node descriptions are in Table 12.4. Figure 12.5 shows the relationship between cost and threshold value. The lowest cost was achieved with the threshold set to 0.25.

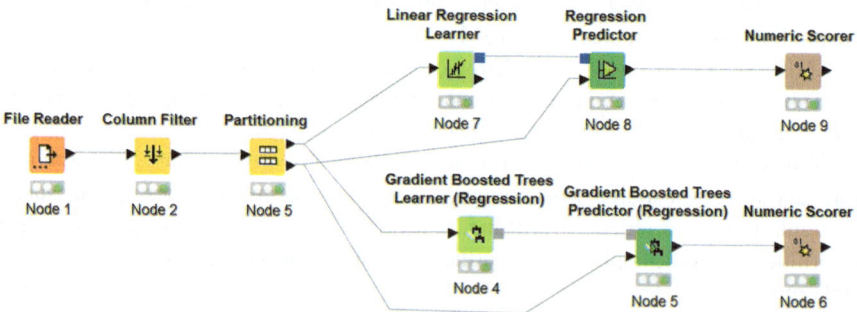

Fig. 12.3 Workflow for comparison of OLS with Gradient Boosted Trees

---

[1] https://www.kaggle.com/datasets/tolgahancepel/toyota-corolla

Table 12.1 Node descriptions for the Fig. 12.3 workflow

Nodes	Descriptions	Input and output ports
Node 1 File reader	Read file: ToyotaCorolla.csv.	<u>Output port</u>: Data table with 1436 rows and 37 columns.

Node 2
Column filter
	Remove selected columns from the input data table.
	    Exclude: Id, Model, Mfg_Year
	    (redundant with Age_08_04).
		<u>Input port</u>: Data table from the output port of Node 1
		<u>Output port</u>: Data table with 1436 rows and 34 columns.

Node 3
Partitioning
	First partition:
	    Select relative: 70%.
	    Select: Draw randomly.
	    Check: Use random seed: 123.
		<u>Input port</u>: Data table from the output port of Node 2
		<u>Upper output port</u>: Training data table with 1005 rows and 34 columns
		<u>Lower output port</u>: Test data table with 431 rows and 34 columns

Node 4
Gradient boosted trees learner (regression)
	Options:
	    Target column: Price.
	    Include: All other columns.
	    Check: Limit number of levels
	    (tree depth): 4.
	Advanced options:
	    Check: Use midpoint splits for
	    numeric attributes.
	    Check: Use binary splits for
	    nominal columns.
		<u>Input port</u>: Data table from the upper output port of Node 3.
		<u>Model output port</u>: Gradient Boosting model

Node 5 Gradient Boosted Trees Predictor (regression)	Apply regression from the Gradient Boosted Trees model to the test data.	<u>Model input port</u>: Model from the model output port of Node 4. <u>Lower input port</u>: Data table from the upper output port of Node 3. <u>Output port</u>: Test data plus prediction with 431 rows and 35 columns.
Node 6 Numeric scorer	Compute statistics with continuous target and prediction.	<u>Input port</u>: Data table from the output port of Node 5. <u>Output port</u>: Prediction statistics with test data

Node 7
Linear regression learner
	Perform linear regression with training data.
	Settings:
	    Target: Price.
	    Include: All variables except
	    Price.
		<u>Input port</u>: Data table from the upper output port of Node 3.
		<u>Model output port</u>: Regression model.

(continued)

Table 12.1 (continued)

Nodes	Descriptions	Input and output ports
Node 8 Regression predictor	Predict Price using regression model and test data.	Upper model input: Model output port from Node 7. Lower input port: Data table from the lower output port of Node 3. Output port: Test data plus prediction with 431 rows and 35 columns
Node 9 Numeric scorer	Compute statistics with continuous target and prediction.	Input port: Data table from the output port of Node 8. Output port: Prediction statistics with test data.

Table 12.2 Prediction accuracy of OLS and Gradient Boosted Trees using the Toyota Corolla data set

Metric	OLS	Gradient Boosted Trees
MAE	851.65	775.02
RMSE	1243.73	1075.43
MAPE	0.08	0.07

Table 12.3 Assumed costs of errors

Actual	Predicted churn	
	no	churn
no churn	0	1 unit
churn	100 units	0

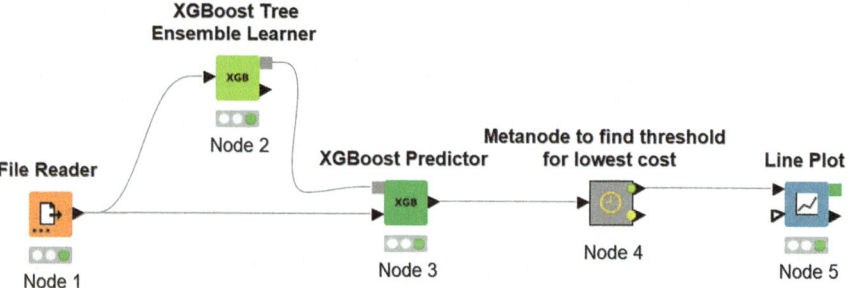

Fig. 12.4 Workflow to find threshold that yields lowest cost

Table 12.4 Node descriptions for the Fig. 12.4 workflow

Nodes	Descriptions	Input and output ports
Node 1 File reader	Read file: ChurnData.csv.	<u>Output port:</u> Data table with file kicked cars file table with 7043 rows and 20 columns.
Node 2 XGBoost tree learner	Create a tree-based XGBoost model for classification. Options: 　Target column: Churn. 　Include: All columns except churn 　Check use static random seed: 123. 　Leave other settings on the page to default. 　Booster page: 　Leave settings on the page to default.	<u>Input port:</u> Data table from the output port of Node 1. <u>Model output port:</u> XGBoost model. <u>Lower output port:</u> XGBoost feature importance (not used).
Node 3 XGBoost predictor	Classifies the churn data using XGBoost model.	<u>Model input port:</u> Model output port from Node 2. <u>Lower input port:</u> Data table from the upper output port of Node 3. <u>Output port:</u> Data table with 7043 rows and 23 columns.
Node 4 Metanode to find threshold for lowest cost	Produces accuracy statistics and a confusion matrix for the XGBoost model training set. Scorer: 　First column: IsBadBuy. 　Second column: Prediction (IsBadBuy).	<u>Input port:</u> Data table from the output port of Node 3. <u>Upper output port:</u> Data table with expected costs and threshold values. <u>Lower output port:</u> Data table with optimum threshold and lowest cost.
	Node M4-1 interval loop start Threshold settings: From 0.1 to .6 in increments of .01.	
	Node M4-2 variable to table column Adds the threshold from the loop to the input data	
	Node M4-3 column expressions 　Computes the cost: 　If churn = no and prediction yes, then cost = 1; 　If churn = yes and prediction = no, then cost = 50.	
	Node M4-4 math formula 　Computes the sum of the costs for all cases.	
	Node M-5 row filter 　Filters all but first row which contains expected cost.	

(continued)

Table 12.4   (continued)

Nodes	Descriptions	Input and output ports
	Node M-6 loop end Cumulates the expected cost for each threshold value.	
	Node M-7 sorter Sorts the output of M_6 from lowest to highest.	
	Node M-8 row filter Retains the first row of the sorted output with the lowest cost.	
ᵗ	Node M-9Column filter Retains only the expected cost and threshold value.	
Node 5 Line plot	Options: Choose column for x-axis: threshold_value nclude: Expected value 1. Label for x axis: Threshold. Lablel for y axis expected cost.	Input port: Data table from the output port of Node 4.

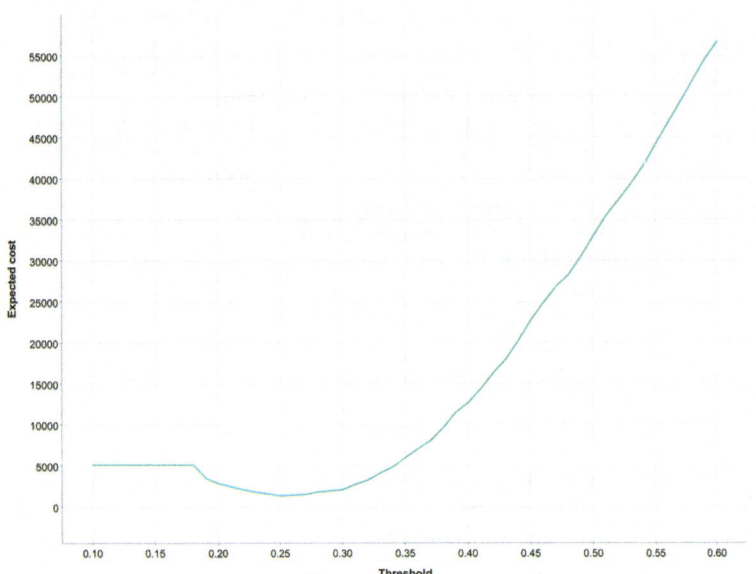

Fig. 12.5   Expected cost versus threshold value

12.5 Summary

Ensemble models combine multiple diverse models to improve prediction accuracy and stability. Popular techniques include bagging, random forests, boosting, and XGBoost. Ensemble modeling has become quite popular since the 1990s fast computers and efficient programs are available to accomplish the needed multiple estimates).

The ensemble models are generally useful but have two main problems. First, while they are very powerful, they can fit a model to the Training data, which does not generalize to the Validation and Test data. Second, ensemble models such as XGBoost do not produce a nice tree like a basic decision tree. So, interpretation and communication of results are more challenging. The ensemble models using decision trees can create hundreds of individual trees, so there is no way to create a visual chart showing the resulting tree model.

References

Adebayo, S. (2020). *How to solve a problem.* https://dataaspirant.com/author/samuel-adebayo/. Accessed 29 July 2023.
Breiman, L. (1966). Bagging predictors. *Machine Learning., 24,* 123–140.
Brownlee, J. (2020). *Why use ensemble learning?* https://machinelearningmastery.com/why-use-ensemble-learning/. Accessed 29 July 2023.
Chen, T., & Guestrin, C. (1996). XGBoost: A scalable tree boosting system. In *Proceedings of the 22nd ACM SIGKDD international conference on knowledge discovery and data mining* (pp. 785–94). https://doi.org/10.1145/2939672.2939785. Accessed 29 July 2023.
Freund, Y, & Schapire, R. E. (1996). Experiments with a new boosting algorithm. In *Machine learning: Proceedings of the thirteenth international conference.* https://cseweb.ucsd.edu//yfreund/papers/boostingexperiments.pdf. Accessed 29 July 2023.
Friedman, J. (2001). Greedy function approximation: A gradient boosting machine. *Annals of Statistics., 20*(5), 1189–1232.
Friedman, J. (2002). Stochastic gradient boosting. *Computational Statistics & Data Analysis., 38*(4), 367–378.
Ho, T. K. (1998). The random subspace method for constructing decision forests. *IEEE Transactions on Pattern Analysis and Machine Intelligence., 20*(8), 832–844.
Seni, G., & Elder, J. F. (2010). *Ensemble methods in data mining: Improving accuracy through combining predictions.* Morgan & Claypool Publishers.
Surowiecki, J. (2005). *The wisdom of crowds: Why the many are smarter than the few and how collective wisdom shapes business, economies.* Societies and Nations. Anchor Books.
Wallis, K. F. (2014). Revisiting Francis Galton's forecasting competition. *Statistical Science., 29*(3), 420–424.

Chapter 13
Cluster Analysis

Cluster analysis is a set of methods for identifying groups of observations using a measure of proximity. There are likely hundreds of different clustering methods available that have been designed for specific problems and data types. This chapter will focus on a subset of methods available in KNIME.

Clustering finds groups of objects (records, people, items, documents) such that the objects within each group are similar in some sense to one another and distinct from the objects in other groups. For example, if three clusters, denoted by A with 10 objects, B with 20 objects, and C with 15 objects, are formed from a set of 45 observations, then the 10 objects in cluster A should be like one another according to a similarity measure. The same should be true of clusters B and C. However, the observations in cluster A are quite distinct (using the same measure) from those in cluster B. This should likewise hold for clusters A versus C and clusters B versus C.

13.1 How Many Clusters Are There?

Deciding on the number and composition of clusters is not always clear. Consider the simple example in Fig. 13.1 of ten points in two dimensions. By visual inspection, one could conclude that there are three, four, or five clusters.

Cluster analysis is an "automatic" technique that does not require a defined target and thus is considered an unsupervised model. Clustering is also a descriptive technique; it is up to the analyst to decide whether the number and composition of the clusters are interesting and relevant to the problem at hand. Cluster analysis can be an intermediate step in an analytics workflow, with the clusters formed becoming inputs to other models. Since it is an unsupervised model, clustering, by itself, is not useful for prediction. However, once clusters are formed, cluster membership can be used as a categorical target along with predictor variables in a predictive model.

© The Author(s), under exclusive license to Springer Nature Switzerland AG 2023
F. Acito, *Predictive Analytics with KNIME*,
https://doi.org/10.1007/978-3-031-45630-5_13

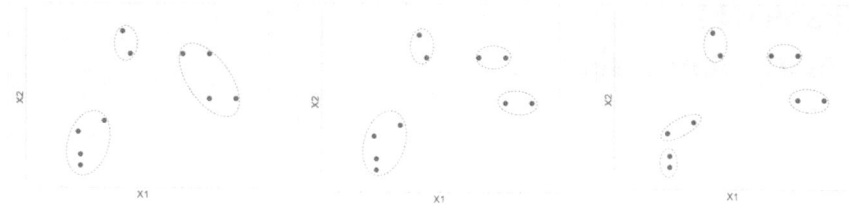

Fig. 13.1 Illustration of how the number of clusters can be ambiguous

Table 13.1 The number of possible clusters by the number of objects	# of objects	# possible clusters
	3	5
	4	15
	5	52
	6	203
	7	877
	8	4140
	9	21,147
	10	115,975

Clustering is used in many scientific and applied fields where the goal is to simplify, condense, or classify situations – to take many data points and somehow extract the essential groups or segments from them.[1] No a priori assumptions are needed to make regarding the number of groups or their structure; the goal is usually one of discovery and description. Sometimes the objective is to find "natural groups" within a data structure. Discovering natural groups is not always easy because of the huge number of possible ways that groups can be formed (and well-defined groups may not exist). One approach might be to investigate all possible groupings and decide which is the "best" using a pre-defined criterion. Table 13.1 shows how the number of possible clusters increases with the number of objects in a data set. The number of possible clusters dramatically increases with the number of objects. Even a moderate number of observations would make the problem computationally unfeasible.

---

[1] Cluster analysis goes by different names in various disciplines and includes numerical taxonomy, pattern recognition, typology, and clumping.

13.2 Recommended Steps in Running a Cluster Analysis

As discussed in the previous section, there are many ways to cluster observations in a data set. While there is no established "cookbook" for obtaining a valid clustering result, steps can be taken to increase the likelihood of a useful result. These steps are discussed in the following sections:

1. Clearly state the objective of the analysis.
2. Select the variables to describe the objects to be clustered.
3. Check that the data set contains clusters.
4. Select a proximity measure.
5. Select a clustering algorithm.
6. Validate the clustering results.
7. Describe the clusters.

Step 1: Clearly State the Objective of the Analysis
The first step is to establish the goal of a cluster analysis application. It might be to form segments of customers for marketing purposes. Or it might be to identify anomalous observations in a fraud detection project. Another purpose is to create a taxonomy or typology to understand medical conditions further. Clustering could also be used to simplify a huge data set into smaller subgroups that could then be used for further analysis. A clustering result can be used as a categorical target variable for predictive analysis. Finally, clustering is sometimes used as an exploratory technique to understand a data set better. Whatever the reason for clustering a data set, it is important first to decide the goal since this will help inform other decisions needed to run the analysis.

Step 2: Select the Variables to Describe the Objects to Be Clustered
This is a critical step since it will determine the analysis results. Depending on the goal of the clustering, it may be possible to form hypotheses about which variables to include. For example, in market segmentation, characteristics of customers such as age, purchase history, income, and education might be relevant. If the goal is to find groups of stocks with similar financial profiles, then the variables could be ratios such as inventory to sales, income to sales, debt to assets, etc.

The type of variable and the measurement scale are also important. In most clustering algorithms, the objects are assumed to be measured on numeric scales, but alternatives exist for categorical variables. Some applications work with a set of binary indicators. More complex situations arise when a mixture of variable types, such as numeric and categorical, in the same analysis, is used.

If certain variables have a much larger range of values or variance than others, these can dominate clusters' formation. So, it is usually recommended that variables be standardized before computing distances. Whether or not standardization should be done depends on domain knowledge in the context of a specific problem.

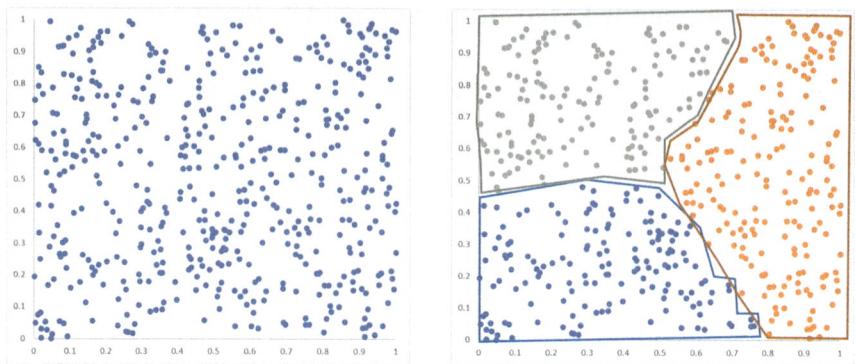

Fig. 13.2 "Clusters" of randomly distributed data

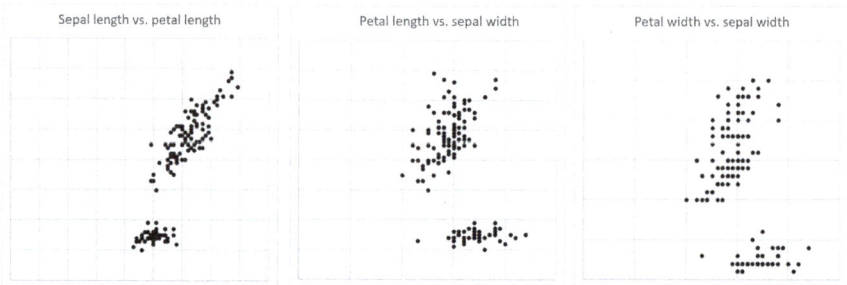

Fig. 13.3 Scatterplots of iris data on sepal and petal dimensions

Step 3: Check That the Data Set Contains Clusters

Since cluster analysis is an unsupervised technique, it is possible to "discover" clusters even in random data distributions. This has been called the "Clustering Illusion" (Kapri, 2019). The left panel in Fig. 13.2 shows a scatterplot with a uniform distribution of 500 observations in two dimensions. The random data was submitted to a clustering program and three clusters were specified. The result is shown in the right panel of Fig. 13.2. Despite the randomness of the data, three "clusters" of sizes 252, 256, and 492 were found. Rerunning the cluster analysis would result in a totally different pattern of clusters.

Creating a visualization is one way to assess whether a data set contains meaningful clusters. For example, the scatterplot in Fig. 13.2 does not have well-defined clusters. If there are more than two dimensions, then pairs of variables can be plotted. For example, consider the Iris data set, which has four measures for each type of flower. Three two-way scatterplots in Fig. 13.3 clearly show at least two clusters.

If there are too many variables to create visualizations, principal components analysis can be used to extract scores. Observations can be plotted on two components at a time.

Observations (objects)	Attributes (columns or variables)				
	a_1	a_2	a_3	...	a_p
o_1	x_{11}	x_{12}	x_{13}	...	x_{1p}
o_2	x_{21}	x_{22}	x_{23}	...	x_{2p}
o_3	x_{31}	x_{32}	x_{33}	...	x_{3p}
o_4	x_{41}	x_{42}	x_{43}	...	x_{4p}
o_5	x_{51}	x_{52}	x_{53}	...	x_{5p}
⋮	⋮	⋮	⋮	⋮	⋮
o_n	x_{n1}	x_{n2}	x_{n3}	...	x_{np}

Fig. 13.4 Typical data table structure for cluster analysis

The Hopkins statistic (Lawson & Jurs, 1990) can be used to assess the clustering tendency of a data set by measuring the probability that a uniform data distribution generates a given data set. In other words, it considers the spatial randomness of the data. This test is not available in KNIME, but an implementation in R can be run in KNIME. If the analyst is confident that clusters may exist, then the next step is to determine how to assess the closeness or proximity of observations in the data set since cluster analysis is based on grouping similar objects.

Step 4: Select a Proximity Measure
Cluster analysis is based on similarity, which expresses how close objects are to one another. Dissimilarity (often measured as distance) is the inverse of similarity and indicates how distinct objects are from one another. To cluster objects, it is necessary to define what criterion will be used to group them. Selecting the measure of similarity or distance is a fundamental decision that greatly affects the results. "…Poor choices of the distance function can sometimes be disastrously misleading depending on the application domain" (Aggarwal, 2015).

While the choice of a measure is critical, no flowchart or rules are available to guide the selection, nor is there one best measure. It may be helpful to address these questions: (a) What constitutes distance (or similarity) in the context of the problem being solved? (b) How should similarity be quantified?

The data table in analytics typically consists of an object (or observations) in the rows and attributes (or variables) in columns, as shown in Fig. 13.4.

For the present discussion, assume that a dissimilarity or distance matrix is calculated from the data table in Fig. 13.4. In hierarchical clustering, distances must be computed between each pair of observations in a data set. This creates a square matrix of distances that has $(n \times (n - 1))/2$ entries plus zeros along the diagonal.[2]

---

[2] If a similarity measure is used, an $(n \times (n-1))/2$ matrix with ones along the diagonal is created.

Objects (in rows)	Attributes (in columns)				
	a_1	a_2	a_3	...	a_k
a_1	0				
a_2	d_{21}	0			
a_3	d_{31}	d_{32}	0		
⋮	⋮	⋮	⋮	⋮	
a_k	d_{k1}	d_{k2}	dk_3	...	0

Fig. 13.5 k by k dissimilarity matrix for cluster analysis

Figure 13.5 shows the structure of a dissimilarity matrix. Each entry in the dissimilarity proximity matrix measures the dissimilarity of a pair of objects in the data table. For example, entry d_{21} in Fig. 13.5 contains the dissimilarity measure between rows o_2 and o_1 in the 13.4 data table. The diagonal elements of the proximity measure are all zeros since each element is equivalent to itself. Only the lower half triangle of the matrix is shown in Fig. 13.5 since the matrix is symmetric.

Dissimilarity or distance can be measured in different ways. Harmouch (2021), for example, explains 17 different methods. In KNIME, the Distance Matrix Calculate node provides the choice of the six metrics, four of which will be described in the following sections: Euclidean, Manhattan, Tanimoto, and Cosine measures. Implementing k-means in KNIME only uses Euclidean distance, but the R package 'ClusterR' offers a variety of distance metrics.

Euclidean Distance for Quantitative Features
The most common dissimilarity measure used in clustering is Euclidean distance. To compute Euclidean distance among objects measured on p attributes, the sum of the squared differences between each pair of objects (the o_i's from 13.4) is computed. In certain situations, the analyst may want to weigh some of the features more heavily than others in calculating Euclidean distance, and this can be done using the a_i weights, as shown in Eq. 13.1.

$$d\left(r_i, r_j\right) = \sqrt{\sum_p^{t=1} a_i \left(r_{i,t} - r_{j,t}\right)^2} \tag{13.1}$$

Where $d(r_i, r_j)$ = the distance between object r_i and r_j,
p = number of attributes, t represents a value on an attribute, and a_i is a user-defined weight.
The a_i are usually set equal to 1.0 unless the analyst believes some attributes are more important in defining distance.

Table 13.2 Hypothetical bank loan customers

Features	Customers		
	A	B	C
Age	28	55	28
Years of education	12	18	16
Credit rating	680	750	690
Annual income	35,000	45,000	45,000

Table 13.3 Euclidean distances among customers

	Customers		
	A	B	C
A	00.0	–	–
B	10,000.0	00.0	–
C	10,000.0	66.0	00.0

Table 13.4 Standardized customer features

Features	Customers		
	A	B	C
Age	−0.707	1.414	−0.707
Years of education	−0.707	−4.950	−0.707
Credit rating	−0.863	7.348	0.000
Annual income	−1.411	0.707	0.707

Before using Euclidean distance in a cluster analysis, it is always necessary to consider the measurement scales used for the variables unless the analyst specifically desires that the large magnitude features are more critical in defining distance. Variables with comparatively large magnitudes will likely affect the distance measures more. For example, consider the three hypothetical customers applying for a bank loan shown in Table 13.2. A and B are very close except on income, while customers B and C have the same income.

Euclidean distances among customers A, B, and C are shown in Table 13.3. The distances indicate that customers B and C are by far the most similar. The distances are almost completely dependent on income.

If the customer characteristics are standardized to zero mean and unit standard deviation, the features appear as in Table 13.4, and the Euclidean distances are now shown in Table 13.5.

By standardizing the features, customers A and C are shown to be closer than customers B and C, which makes more sense.

Another more subtle problem occurs when there are attributes that are highly correlated with one another. For example, measures such as horsepower, engine displacement, weight, and length might be used to form clusters of automobiles

Table 13.5 Euclidean distances using standardized features

	Customers		
	A	B	C
A	00.0	–	–
B	9.717	0.000	–
C	2.290	8.746	0.000

based on their physical characteristics. These variables are likely to be correlated, and highly correlated variables will have more weight in measuring distance. One remedy is to run principal components on the attributes to form uncorrelated measures before clustering. The analysis proceeds in two stages: apply principal components to a correlation matrix of the raw data and then submit the component scores to cluster analysis.

Manhattan Distance
Manhattan distance captures the distance among rows of data by aggregating the absolute differences in values of the attributes. The formula for Manhattan distance is shown in Eq. 13.2:

$$d\left(r_i, r_j\right) = \sqrt{\sum_p^{t=1} a_i \left| r_{i,t} - r_{j,t} \right|} \tag{13.2}$$

The same comments about standardizing data before clustering discussed with Euclidean distances apply to Manhattan distances.

Results obtained using Euclidean versus Manhattan distances should give similar results in most cases. However, if there are a few observations with extreme values, Manhattan distance was found to be a better alternative to Euclidean distance when the number of dimensions, p, is 20 or more (Aggarwal et al., 2001).

Cosine Similarity
A matrix with observations in rows and variables in columns is the most common way of representing a data set. Multivariate data can also be structured as vectors with a direction and length. The cosine of the angle between two vectors measures similarity. When the two vectors point in the same direction, the cosine is 1.0, but if the two vectors are at right angles, the cosine is 0.0, and if the vectors are exactly opposite, the cosine is -1.0. Consider the following data table with three observations and two variables shown in Table 13.6. This data can be converted into two vectors in three dimensions shown, one dimension for each observation, in Fig.13.6.

Equation 13.3 shows the formula for computing cosine similarity.

$$cosine\ similarity = \frac{\sum_i^n x_i \times y_i}{\sqrt{\sum_i^n x_i^2}\ \sqrt{\sum_i^n y_i^2}} \tag{13.3}$$

Where x_i and y_i are components of the vectors X and Y
and n = the number of observations.

Table 13.6 A simple
data table

Obs	X	Y
1	5	8
2	5	1
3	1	5

Fig. 13.6 Data vectors
from Table 13.6 in 3
dimensions

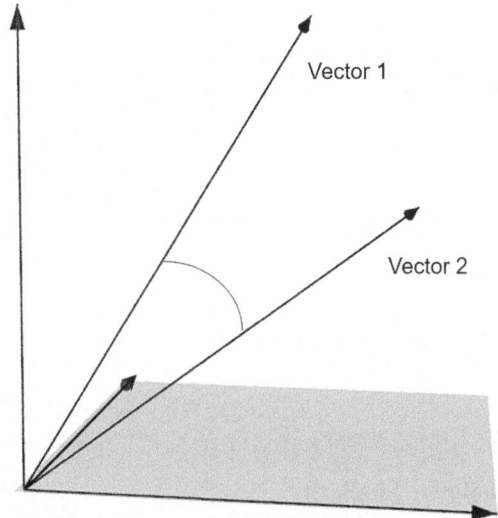

Table 13.7 Word frequencies in automobile descriptions

Autos	quick	mileage	brakes	handling	excellent	style	comfort	quiet	value	economy
Auto 1	5	0	3	0	2	0	0	2	0	0
Auto 2	3	0	2	0	1	1	0	1	0	1
Auto 3	0	7	0	2	1	0	0	3	0	0
Auto 4	10	0	6	0	4	0	0	4	0	0

Cosine distance is useful in text analysis where the frequency of terms is placed in multiple vectors and image comparison. For example, consider the four sets of words derived from descriptions of automobiles shown in Table 13.7.

Applying Eq. 13.3 to the rows in the table yields the similarity matrix in Table 13.8.

This shows that Auto 1's description had much in common with Auto 2's description but that Auto 3 differed from Auto 1 and Auto 2. Auto 4, despite having exactly twice the frequencies of Auto 1, is still scored as identical to Auto 1. Unlike Euclidean distance, cosine similarity assesses the overall pattern of characteristics.

Since some clustering techniques require distances or dissimilarity measures rather than similarity metrics, it may be necessary to convert cosine similarity to a

Table 13.8 Example similarity matrix

	Auto 1	Auto 2	Auto 3	Auto 4
Auto 1	1.000			
Auto 2	0.940	1.000		
Auto 3	0.160	0.120	1.000	
Auto 4	1.000	0.940	0.160	1.000

Table 13.9 Example of a binary data matrix

Name	Test1	Test2	Test3	Test4	Test5	Test6	Test7	Test8	Test9
Smith	1	1	1	0	1	1	0	0	0
Jones	1	1	1	0	1	0	0	0	1
Baker	1	0	1	1	0	1	1	0	0

1 = medical test performed, 0 = medical test not performed

distance. A common way to convert similarity to distance is to use Eq. 13.4 for normalized vectors A and B.

$$Distance(A,B) = 2 - cosine\ similarity(A,B) \qquad (13.4)$$

Tanimoto Similarity

Tanimoto similarity is one measures useful for comparing binary vectors. The measure is calculated by enumerating binary values in pairs of records.

$$Tanimoto\ similarity = \frac{d}{b+c+d} \qquad (13.5)$$

Where d = the number of columns where record A and B both have 1's,
b = the number of columns where record A has a 1 and B has a 0,
c = the number of columns where record B has a 1 and A has a 0.

Note that columns where both records have zeros are not counted. As an illustration, consider comparing medical tests performed on three patients. The hypothetical data is shown in Table 13.9. If two patients have had a specific test, this contributes to similarity, but if both did not, the similarity is not increased.

For Smith and Jones, $d = 4$, $b = 1$, and $c = 1$. So, the Tanimoto similarities are computed as follows:

$For\ Smith\ and\ Jones: d = 4, b = 1, and\ c = 1, so\ Tanimoto\ similarity = \dfrac{4}{1+1+4} =$
0.67

$$For\ Smith\ and\ Baker : d = 3, b = 2, and\ c = 2, so\ Tanimoto\ similarity = \frac{3}{2+2+3} =$$

0.43

$$For\ Jones\ and\ Baker : d = 2, b = 3, and\ c = 3, so\ Tanimoto\ similarity = \frac{2}{2+3+3} =$$

0.25

The range of the Tanimoto coefficient is 0.0–1.0, so to convert to a distance measure, use Eq. 13.6.

$$Distance(A,B) = 1 - Tanimoto\ similarity(A,B) \tag{13.6}$$

Step 5 Select a Clustering Algorithm
Each clustering method begins with a data set of n observations on p attributes or dimensions. The methods differ in terms of the criterion used to form clusters, the ability to work with diverse types of data, computer memory requirements, and processing speed. The following, adapted and modified from (Han et al., 2012), lists four major types of cluster analysis:

- Hierarchical clustering

    - Builds a hierarchy of 1 to n clusters (where n is the number of objects) so the number of clusters does not have to be pre-specified.
    - Uses a $p \times p$ matrix of proximities based on distance, correlation, binary, matching, and other measures.

- k means partitioning

    - Produces a mutually exclusive, pre-specified number of clusters of spherical shape.
    - Uses distances based on k-means, k-medoids, and k-modes.

- Density-based clustering

    - Defines arbitrarily shaped clusters as dense regions of objects separated by sparse regions.
    - Tends to filter out noise and outliers.

- Fuzzy (soft) clustering

    - Requires a pre-defined number of clusters.
    - Assigns objects as probabilities of belonging so that each object can be associated with more than one cluster.

13.3   Hierarchical Cluster Analysis

Hierarchical clustering refers to a set of related techniques with software implementations offering alternative settings and choices of metrics. Some software tools limit the choices to a small number, while others offer an almost bewildering list of different approaches. (KNIME is intermediate in the number of options provided, but many other types of hierarchical clustering are available in R and Python that can be used with KNIME.)

The hierarchical clustering process can proceed either bottom-up or top-down. Bottom-up, or agglomerative clustering, starts with all objects in separate clusters and iteratively creates clusters until all objects are in a single cluster. Top-down, or divisive clustering, starts with all objects in a single cluster and recursively splits nodes until a stopping criterion is reached. The agglomerative process is the most common and is included in KNIME. The divisive method requires exploration of $2^{n-1}-1$ possible ways to split the initial cluster, which can lead to impossibly long computational times. For example, with 25 objects, divisive clustering requires trying nearly 17 million ways to split the first cluster.[3]

A simple conceptual example of hierarchical clustering is used to demonstrate how hierarchical clustering works. For this example, assume that nine objects are measured on two dimensions, as shown in Table 13.10. Both x_1 and x_2 have been standardized to zero mean and unit standard deviation.

Euclidean distances were computed for the 9 objects in the rows of Table 13.10. Since there are nine objects, $(9 \times 8)/2 = 36$ distances are needed, which are listed in Table 13.11.

The hierarchical sequence of clustering is shown in Fig. 13.7. (The order of the columns was changed to clarify the diagram.) To begin, there are nine separate clusters, each with one object. The 36 distances are ranked from smallest to largest, and

Table 13.10 Simulated data for demonstration of hierarchical cluster analysis

Observation	x_1	x_2
A	2.05	0.93
B	0.70	1.28
C	1.17	1.63
D	−0.45	−0.89
E	−0.01	−0.86
F	−0.59	−1.59
G	−1.03	−0.46
H	−0.94	−0.13
I	−1.18	−0.46

---

[3] There is an algorithm called DIANA available in R and Python which uses a heuristic method of reducing the number of splits required with divisive clustering (Kaufman and Rousseeuw 2005).

Table 13.11   Example similarity matrix

	A	B	C	D	E	F	G	H	I
A	0.000								
B	1.395	0.000							
C	1.124	0.586	0.000						
D	3.092	2.456	2.996	0.000					
E	2.729	2.255	2.755	0.441	0.000				
F	3.650	3.147	3.670	0.714	0.932	0.000			
G	3.379	2.454	3.034	0.722	1.096	1.213	0.000		
H	3.172	2.163	2.748	0.904	1.182	1.501	0.342	0.000	
I	3.516	2.562	3.145	0.847	1.236	1.275	0.150	0.408	0.000

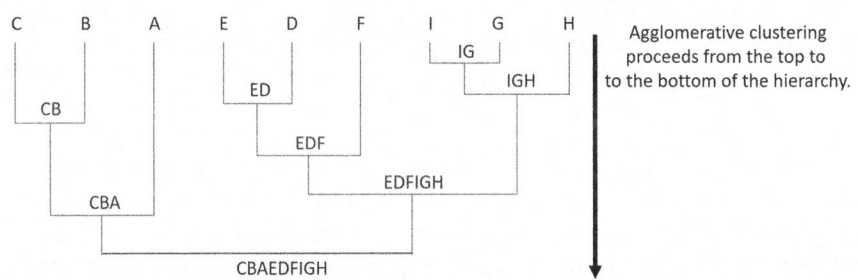

Fig. 13.7   Dendrogram for hierarchical clustering example

the object pair with the smallest distance is joined, forming the first cluster, which contains objects I and G.

At this point, there are eight clusters: C, B, A, E, D, F, IG, H. Distances are again computed, and object H is joined with IG forming cluster IGH forming seven clusters. Next, six clusters are formed: C, B, A, ED, F, IGH. The process continues until all the objects are in a single cluster. Note that the number of clusters need not be specified in advance. Various metrics for selecting the number of clusters are discussed later. The hierarchical nature of this approach is evident from the fact that once a cluster is formed, the algorithm does not go back to check if that cluster is still optimum.

The distances among clusters can be computed unambiguously only at the very first stage of the analysis when all objects are in separate clusters. The question of assessing the distances among clusters arises from the second level down. KNIME provides three measures, called linkages, for determining the distances among clusters: single linkage, complete linkage, and average linkage. In the previous example, single linkage was selected. The results using any of the three linkages will be similar with such a small data set. In larger problems, the choice of linkage metric can result in very different clusters.

The three linkage methods can be illustrated graphically using 25 simulated observations in two dimensions, x_1 and x_2. The scenario assumes that hierarchical clustering has been applied to the point where three clusters have been identified. The question at this stage is which of the three clusters should be joined next.

Single linkage, also called "nearest neighbor," defines the distance between two clusters as the smallest distance between any two points in the clusters. Single linkages are shown in Fig. 13.8 with distances from point F to point H (cluster #1 to cluster #2), N to T (cluster #2 to cluster #3), and S to G (cluster #3 to cluster #1). The shortest distance is N to T, so these two clusters would be joined next. Single linkage is known to produce long, stringy clusters.

Complete linkage, also called "farthest neighbor," takes a more conservative approach. This method determines the largest distance between any points in a pair of clusters. The pair with the "shortest maximum distance" are joined. Figure 13.9 shows the same data with complete linkages. The linkages are point B to point R (clusters #1 to #2), K to W (clusters #2 to #3), and W to A (clusters #3 to #1). In this case, clusters #1 and #2 have the shortest distance based on the farthest points.

Average linkage uses the mean of the distances among all pairs of points in each cluster. Figure 13.10 illustrates the idea, but only a small portion of the distances are depicted to avoid cluttering the diagram. The connections between object A in cluster #1 and each of the seven objects in cluster #3 are shown. Since cluster #1 contains seven objects and cluster #3 contains seven, 49 distances would need to be calculated. The average of the forty-nine interpoint distances becomes the distance between the two clusters. Similarly, all the pairwise distances between clusters #1

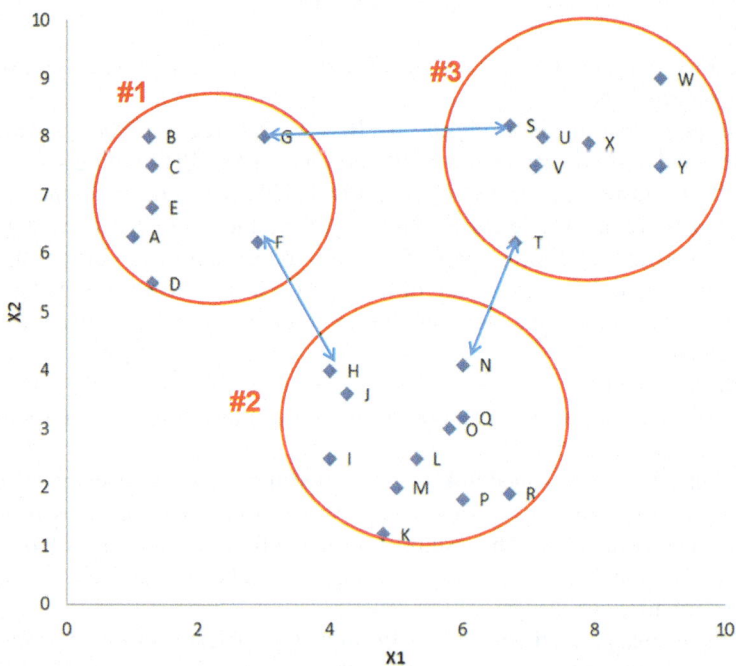

Fig. 13.8 Illustration of single linkage clustering

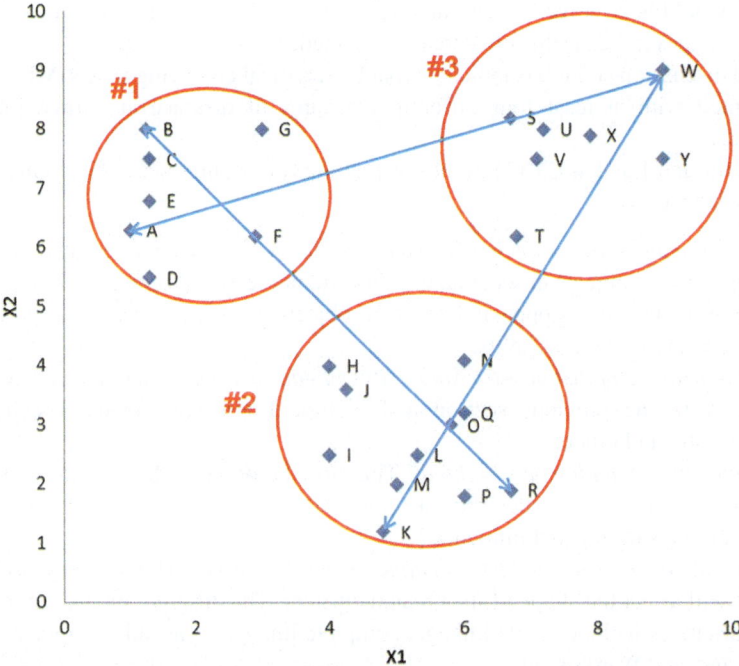

Fig. 13.9 Illustration of complete linkage clustering

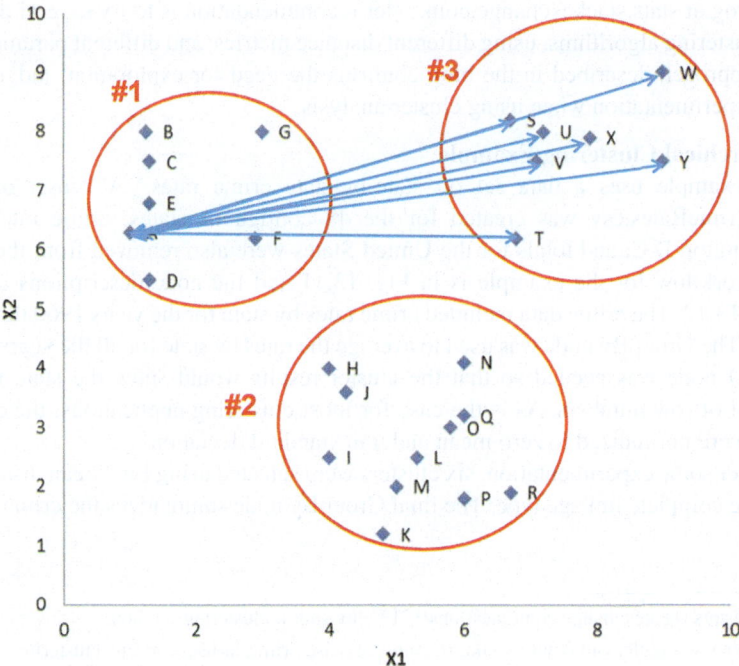

Fig. 13.10 Illustration of average linkage clustering

and #2 would be averaged as the same for clusters #2 and #3. The minimum of the three averages indicates the clusters being joined.

The discussion so far has indicated that hierarchical clustering in KNIME can be performed with at least four different measures of distance and three linkage methods.

Everitt and Landau (2011) point out that analysts using hierarchical clustering face these questions:

- *Should agglomerative or divisive measures be used?* The agglomerative method starts with all objects in separate clusters and proceeds to form groups. The divisive approach begins oppositely with all objects in a single cluster and proceeds by considering various splits.
- *What measure of distance (or similarity) should be used?* Alternatives were discussed in the previous section and include Euclidean, cosine, Manhattan, Tanimoto, and others.
- *Should the variables be weighted?* This may be done so that certain variables have more influence on determining clusters, or should weights be chosen so that all variables have equal influence?
- *What measure of inter-cluster distance should be used?* This is a key decision that will affect the shape of the created clusters. The different inter-cluster rules Alternatives include single linkage, complete linkage, centroid linkage, average linkage, and Ward's method.

As might be expected, there are no definitive answers to these questions. One suggested approach was provided by Mariana Soffer, an analyst Sentisis Intelligence, in a blog at stats.stackexchange.com.[4] Her recommendation is to try several different clustering algorithms, using different distance metrics, and different parameters. The approach described in the blog confirms the need for exploration, judgment, and experimentation when using cluster analysis.

Hierarchical Clustering Example

This example uses a data set on state-by-state crime rates.[5] A subset of file USACrimeRates.csv was created for the 48 contiguous states; crime rates for Washington D.C. and totals for the United States were also removed from the file. The workflow for the example is in Fig. 13.11 and the node descriptions are in Table 13.12. The crime data included crime rates by state for the years 1960 through 2019. The GroupBy node was used to average the rates by state for all the years. The RowID node was needed so that the cluster results would show the state name instead of row numbers. As is the case for most clustering applications, the crime rates were normalized to zero mean and unit standard deviation.

After some experimentation, six clusters were selected using Euclidean distances and the complete linkage type. The final GroupBy node summarizes the crime rates

---

[4] https://stats.stackexchange.com/questions/3713/choosing-a-clustering-method.

[5] https://www.kaggle.com/datasets/vikramamin/statewise-crime-in-usa-k-means-clustering

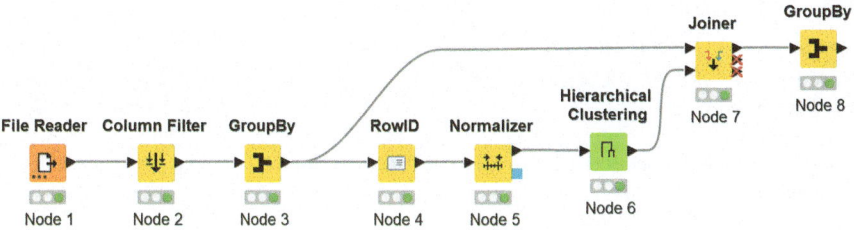

Fig. 13.11 Workflow for hierarchical cluster analysis US crime rates

for six clusters. The cluster results were copied into an Excel file to produce Tables 13.13 and 13.14.

The cluster memberships shown in Table 13.13 show the groups that were formed in terms of similarities of crime rates. There is some geographic correspondence for the states included in each cluster, but there are exceptions as well.

Table 13.14 has the mean crime rates for each cluster. The entries in bold indicate higher than average rates for the respective clusters. Cluster 5 has above average rates for all the crime categories. Cluster 3 is below average in all categories. Cluster 2 has above an average rate for motor vehicle crimes and Cluster 4 is above average only for larceny.

Summary Comments on Hierarchical Clustering
Hierarchical clustering approaches (divisive or agglomerative methods) have several advantages and disadvantages.

Advantages include:

- The results do not depend on the choice of an initial centroid, so re-running the analysis will produce the same results.
- Hierarchical methods can detect clusters of different shapes; it is not restricted to spherical clusters. However, it cannot deal with all the complex shapes that can appear. Density-based clustering, discussed later, can be used in such cases.
- The dendrograms produced are useful graphical techniques for determining the number of clusters.
- Many options can be selected, including different similarity measures.

Disadvantages include:

- The algorithms work best with relatively small numbers of cases since a distance matrix of order n-squared, where n is the number of cases, must be computed. In general, the processing time is of order n-cubed.
- The dendrograms produced with large numbers of cases can be uninterpretable.
- For each step of the clustering process, the following step depends on clusters obtained previously. This may prevent the model from finding the best outcome.

Table 13.12 Node descriptions for Fig. 13.11 workflow

Nodes	Descriptions	Input and output ports
Node 1 File Reader	Read file: USACrimeRates.csv.	Output port: Data table with 2,874 rows by 21 columns.

Node 2
Column Filter
	This node selects the crime rate data from Node 1. identify the clusters.
	Column Filter:
	  Include:
	    State.
	    Data.Rates.Property.Burglary.
	    Data.Rates.Property.Larceny.
	    Data.Rates.Property.Motor.
	    Data.Rates.Violent.Assault.
	    Data.Rates.Violent.Murder.
	    Data.Rates.Violent.Rape.
	    Data.Rates.Violent.Robbery.
	  Exclude: All other columns.
		Input port: Data table from the Output port of Node 1.
		Output port: Data table with 2,874 rows and 8 columns.

Node 3
GroupBy
	This node calculates the average crime rates for all years.
	Groups:
	  Group column(s): State.
	Manual Aggregation:
	  Mean of each crime rate by state.
		Input port: Data file from the Output port of Node 2.
		Output port: Data table with 48 rows and 8 columns.

Node 4
RowID
	Options:
	  Replace RowID.
	  New RowID column: State.
		Input port: Data file from the Output port of Node 3.
		Output port: Data table with 48 rows and 8 columns.

Node 5
Normalizer
	Methods:
	  Include: All continuous variables.
	Settings:
	  Check: Z-Score Normalization
		Input port: Data file from the Output port of Node 4.
		Output port: Data table with 48 rows and 8 columns.

Node 6
Hierarchical
Clustering
	This node adds a new column with cluster membership.
	Options:
	  Number output cluster: 6.
	  Distance function: Euclidean.
	  Linkage type: Complete.
	Include:
	  All continuous variables.
		Input port: Data table from the Output port of Node 5.
		Output port: Data table with 48 rows and 9 columns.

Node 7 Joiner	This node joins the original data with the normalized data so that cluster means can be computed in the original variable metrics. Joiner Settings: Match: State	Input port: Data table from the Output port of Node 6. Output port: Data table with 48 rows and 17 columns.

(continued)

Table 13.12 (continued)

Nodes	Descriptions	Input and output ports
Node 8 GroupBy	This node calculates the average crime rates by clusters. Groups: Group column(s): Cluster. Manual Aggregation: Mean of each crime rate by cluster.	Input port: Data table from the Output port of Node 7. Output port: Data table with 6 rows and 8 columns.

Table 13.13 States in each of six clusters

Cluster 1	Cluster 2	Cluster 3	Cluster 4	Cluster 5	Cluster 6
Illinois	Connecticut	Idaho	Colorado	Arizona	Alabama
Maryland	New Jersey	Iowa	Delaware	California	Arkansas
New York	Massachusetts	Kentucky	Indiana	Florida	Georgia
	Rhode Island	Maine	Kansas	Michigan	Louisiana
		Montana	Minnesota	Nevada	Mississippi
		Nebraska	Ohio	New Mexico	Missouri
		New Hampshire	Oregon		North Carolina
		North Dakota	Utah		Oklahoma
		Pennsylvania	Washington		South Carolina
		South Dakota			Tennessee
		Vermont			Texas
		Virginia			
		West Virginia			
		Wisconsin			
		Wyoming			

Table 13.14 Mean crime rates by cluster

Cluster	Burglary	Larceny	Motor	Assault	Murder	Rape	Robbery
1	890.8	2307.9	**456.1**	**327.4**	**8.3**	27.7	**285.1**
2	839.4	1950.5	**503.8**	195.2	3.5	20.9	125.8
3	579.2	1887.4	182.7	130.3	3.5	22.7	43.3
4	905.7	**2641.1**	330.0	193.7	4.3	33.8	98.7
5	**1259.0**	**2894.2**	**511.8**	**358.6**	**8.4**	**40.9**	**184.2**
6	**999.4**	**2237.8**	311.8	**303.0**	**9.3**	30.8	116.6
Average	912.2	2319.8	382.7	251.4	6.2	29.5	142.3

13.4   k-Means Clustering

Hierarchical clustering described in the previous section is very flexible, allowing various distance measures depending on the data and different linkage rules for joining clusters. However, hierarchical cluster analysis does not scale well because it requires $n(n - 1)/2$ calculations of proximities. Memory and computational requirements become excessive if a clustering problem involves more than a few thousand observations. Partitioning methods, such as k-means, can cluster millions of observations efficiently.

K-means partitioning methods begin with a data set of n observations measured on p continuous variables. These must be converted into continuous measures if nominal variables are to be included. The number of clusters to form, denoted by k, must be set in advance by the analyst. If the analyst wants to explore different numbers of clusters, the procedure must be re-run for each separate value of k.

The algorithm forms groups that are similar in terms of k centroids calculated as multidimensional means of the observations in each cluster. Observations are divided into clusters such that every observation belongs to one and only one cluster; the clusters do not form a tree structure as they do in the hierarchical procedures.

The k-means algorithm begins with k centroids that are usually standardized to prevent those with large means and variances from dominating the results. The k-means algorithm works like this:

1. Initial centroids (or seeds) for k clusters are randomly selected from the observations or first k observations.
2. Assign each observation to the nearest of the k centroids using Euclidean distance to form k clusters.
3. Using the newly formed k clusters, compute new centroids.
4. Using the new centroids from step 3, reassign cases to the nearest new centers again using Euclidean distance to form new clusters.
5. Repeat steps 3 and 4 until the maximum number of iterations (set by the analyst) is reached.

The KNIME documentation for k-means also states that the "… algorithm terminates when the cluster assignments do not change anymore," but the exact criterion is not specified.)[6]

A weakness of k-means is that the results can be very sensitive to the starting centroids.

Determining the Number of Clusters with k-Means

Two popular ways to determine the number of k-means clusters are the "elbow method" and the silhouette coefficient. Both methods require that k-means be run with a sequence of clusters, from two to the highest number expected in the data. The within sum of squares for each number of clusters is plotted versus the number of clusters. An "elbow" in the plot indicates the number of clusters.

The silhouette coefficient gives a quantitative value for the quality of a clustering solution. For each point p, first find the average distance between p and all other points in the same cluster (this is a measure of cohesion, call it A). Then find the average distance between p and all points in the nearest cluster (this is a measure of

---

[6] The KNIME documentation for k-means also states that the "… algorithm terminates when the cluster assignments do not change anymore," but the exact criterion is not specified.

separation from the closest other cluster, call it B). The silhouette coefficient for p is the difference between B and A divided by the greater of the two (max(A,B)).

A recommended way to select the number of clusters is to choose the number with the highest overall mean of the silhouette coefficient (SC) across all cases (Kaufman & Rousseeuw, 2005). If SC > 0.70, the structure of the clusters is strong. If SC is between 0.51 and 0.70, the structure is reasonable. Values lower than 0.5 indicate poor structure.

Experience with the elbow and silhouette coefficient methods indicates that both worked well with clearly defined clusters. The silhouette method provides additional information on the quality of the cluster structure.

Example of k-Means Clustering

A data set was obtained from Kaggle[7] with 200 observations on four variables: gender, age, annual income in $000's, and a "spending score" which ranged from 0 to 100. (The detail behind the spending score was not made available, but it is assumed that a higher score indicates a greater likelihood of spending at the mall.) The first few rows of the data set are shown in Table 13.15.

The workflow to determine the number of clusters using the elbow and silhouette methods is in Fig. 13.12. Descriptions of each node are in Table 13.16.

The within sum-of-squares (Fig. 13.13) and silhouette criteria (Fig. 13.14) both indicated five clusters. A workflow (Fig. 13.15) was created to read the data, normalize it, find five clusters using k-means, and output the final cluster result. Node descriptions are in Table 13.17. The output of the k-means node contains the assigned cluster numbers and the values on the normalized inputs. To insert the variables in the unnormalized form, the output of the k-means node was joined to the original data. This was done to describe the clusters using the original scaling.

The summarized data was copied to an Excel spreadsheet to create the bar charts shown in Fig. 13.16.

Table 13.15 First 10 rows of the Mall Customers data set

RowID	Gender	Age	Income	SPending score
Row0	Male	19	15	39
Row1	Male	21	15	81
Row2	Female	20	16	6
Row3	Female	23	16	77
Row4	Female	31	17	40
Row5	Female	22	17	76
Row6	Female	35	18	6
Row7	Female	23	18	94
Row8	Male	64	19	3
Row9	Female	30	19	72

---

[7] Kaggle data: https://www.kaggle.com/datasets/shwetabh123/mall-customers

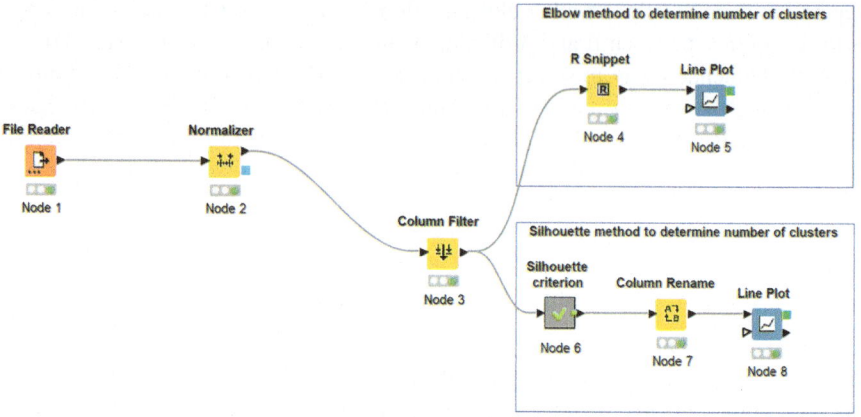

Fig. 13.12 Workflow to determine the number of clusters in the mall data

Summary Comments on k-Means

K-Means and other partitioning methods are among the most used cluster analysis methods. These clustering algorithms have these strengths:

- The complexity (processing time and memory requirements) is linear in the number of objects; therefore, k-means can handle problems with many cases.
- k-means is a useful approach to identifying outlying observations.
- Objects are reassigned as the clustering process iterates and are not fixed as with hierarchical methods.

Weaknesses must also be considered:

- The starting centers are randomly selected, affecting the final cluster compositions. Therefore, with different random initial centers, clustering results may also differ. (One method to deal with this weakness is to restart the clustering process multiple times with different initial seeds to check the stability of the results.)
- The analysts must pre-specify the number of clusters, so the model must be re-run to evaluate various results.
- The algorithm works best with spherical cluster arrays.

The following section discusses density-based clustering, which can deal with arbitrarily shaped clusters.

13.5   Density-Based Clustering

Density-based clustering is an approach that can be used with densely packed points in a multidimensional space. The principal feature of density-based clustering is that arbitrarily shaped clusters can be identified, which may not be identified with

Table 13.16 Node descriptions for Fig. 13.12 workflow

Nodes	Descriptions	Input and output ports
Node 1 File Reader	Read file: Mall_Customers.csv.	<u>Output port:</u> KNIME data table, 200 rows by 4 columns.

Node 2
Column
Filter
	Keep only numerical columns.
	Column Filter:
	  Check Type Selection.
	  Include:
	    Number (double).
	    Number (integer).
	    Number (long).
		<u>Input port:</u> Data table from the Output port of Node 1.
		<u>Output port:</u> KNIME data table, 200 rows by 3 columns.

Node 3
Normalizer
	The three variables had different means and ranges; normalization will avoid weighting cluster membership by mean or range.
	Methods:
	  Include:
	    Age.
	    Annual Income (k$).
	    Spending Score (1–100).
	  Check: Z-Score Normalization.
		<u>Input port:</u> Data table from the Output port of Node 2.
		<u>Output port:</u> Normalized table, 200 rows by 3 columns.

Node 4 R Snippet	This computes the data for determining the number of clusters using the elbow method.	<u>Input port:</u> Data table from the Output port of Node 3. <u>Output port:</u> KNIME data table, 14 rows by 2 columns, with no. of clusters and wss.

	R code:
	<pre>library(factoextra)
	set.seed(5533)
	elbowData <- fviz_
	nbclust((knime.
	in),kmeans,method="wss",k.
	max=15)
	elbowDF<- as.data.
	frame(elbowData$data[-1,])
	colnames(elbowDF) <- c("# of
	clusters", "wss")
	knime.out <- elbowDF</pre>	

Node 5 Line Plot	This creates a line plot of within sum of squares by number of clusters. Options: Choose column for x-axis: # of clusters. Include: wss.	<u>Input port:</u> Data table from the Output port of Node 4.
Node 6 Silhouette criterion	This Metanode loops through clusters from 2 to 15 and calculates the mean silhouette coefficient.	<u>Input port:</u> Data table from the Output port of Node 3. <u>Output port:</u> 14 rows, 2 columns with k=number of clusters and Objective value for each number of clusters.

(continued)

Table 13.16 (continued)

Nodes	Descriptions	Input and output ports

	Node M6_1
	Parameter Optimization Loop Start.
	Standard settings:
	    Parameter: k.
	    Start value: 2.
	    Stop value: 15.
	    Step size: 1.
	    Check Integer.	

	Node M6_2
	K-Means
	    K-Means Properties:
	    Number of clusters: controlled by Node M6_1.
	    Check Random initialization; Use static random seed = 123.
	    Include:
	        Age.
	        Annual Income.
	        Spending score.	

	Node M6_3
	Silhouette Coefficient.
	Settings:
	    Include:
	        Age.
	        Annual Income.
	        Spending Score.
	    Clustering Column Selection: Cluster	

	Node M6_4
	Parameter Optimization Loop End.
	Options:
	    Flow variable with objective function value: Overall
	    Mean Silhouette Coefficient.
	    Function should be: Check maximized.	

Node 7 Column Rename	Rename column "Objective value" to Silhouette coefficient.	<u>Input port:</u> Data table from the Output port of Node 6. <u>Output port:</u> 14 rows, 2 columns with k=number of clusters and Silhouette coefficient.

Node 8
Line Plot
	This creates a line plot of the silhouette coefficient by number of clusters.
	Options:
	    Choose column for x-axis: # of clusters.
	    Include: Silhouette coefficient.
		<u>Input port:</u> Data table from the Output port of Node 7.

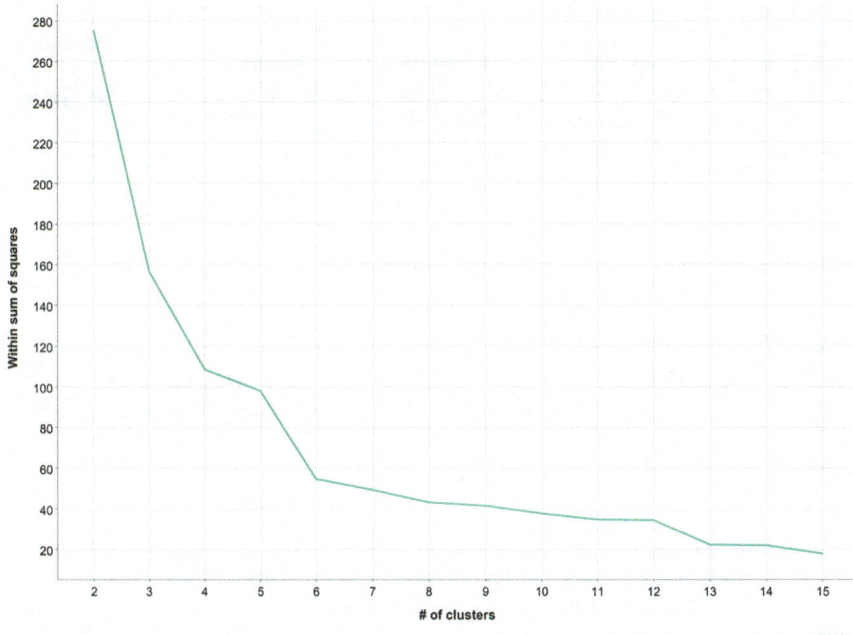

Fig. 13.13 Within sum-of-squares by number of clusters for mall data

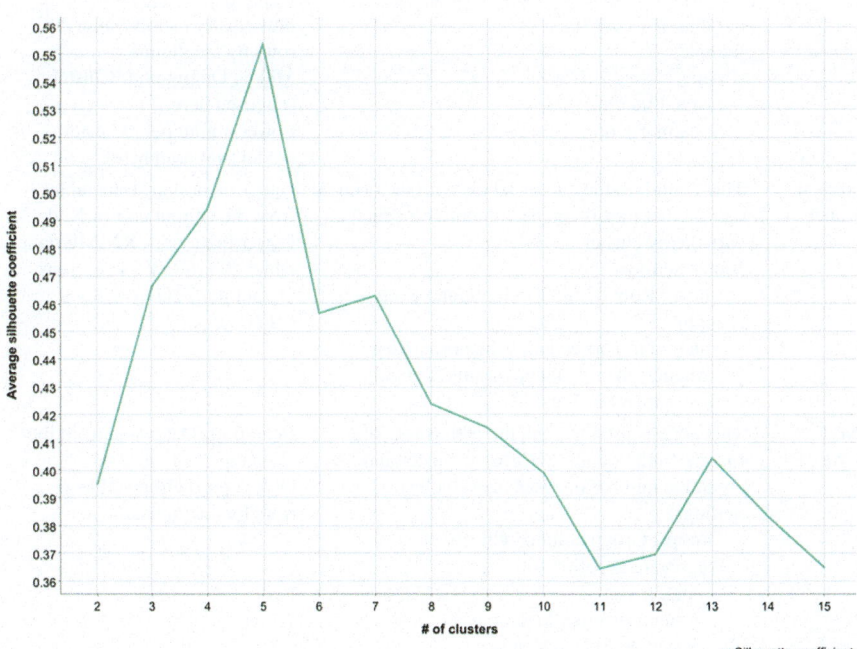

Fig. 13.14 Silhouette coefficient by number of clusters for mall data

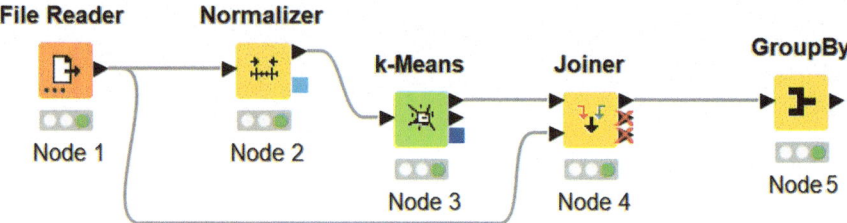

Fig. 13.15 Workflow for k-means clustering of mall data

Table 13.17 Node descriptions for Fig. 13.15 workflow

Nodes	Descriptions	Input and output ports
Node 1 File Reader	Read file: Mall_Customers.csv.	Output port: KNIME data table, 200 rows by 4 columns.
Node 2 Normalizer	The three variables had different means and ranges; normalization will avoid weighting cluster membership by mean or range. Methods: Include: Age, Annual Income (k$), Spending Score (1–100). Check: Z-Score Normalization.	Input port: Data table from the Output port of Node 1. Output port: Normalized table, 200 rows by 4 columns.
Node 3 k-Means	K-Means Properties: Number of clusters was set to 4. Check Random initialization; Use static random seed = 0. Include: Age. Annual Income. Spending score.	Input port: Data table from the Output port of Node 2. Upper output port: Input data plus cluster membership; 200 rows by 5 columns. Middle Output port: Clusters (not used here). Lower output port: Cluster model (not used here).
Node 4 Joiner	This combines the cluster labels with the original data so that means can be computed for each variable by cluster. Joiner Settings: Match: Row ID for Top Input and Bottom Input. The output port contains the normalized variables, the raw input variables, and the cluster memberships.	Upper input port: Data table from the Output port of Node 3. Lower input port: KNIME data table, 200 rows by 4 columns. Output port: 200 rows by 9 columns.
Node 5 GroupBy	This node is used to compute means for Age, Income, and Spending Score by cluster and count the number of cases in each cluster. Settings: Groups: Group column: Cluster. Aggregate as follows: Age: Mean. Annual Income: Mean. Spending Score: Mean. Also: Age: Count.	Input port: Data table from the Output port of Node 4. Output port: Group table, 6 rows by 4 columns.

Cluster #	Number of cases	Percentage of cases	Age (yrs)	Age vs. average	Income ($1,000s)	Income vs average	Spending (0 - 100)	Spending vs. average
1	23	11.5%	25.5		26.3		78.6	
2	57	28.5%	55.5		47.9		41.9	
3	49	24.5%	27.1		52.0		41.0	
4	40	20.0%	32.9		86.1		81.5	
5	31	15.5%	44.4		89.8		18.5	

Fig. 13.16 Cluster descriptions for mall data

k-means or hierarchical approaches. DBSCAN (Ester et al., 1996) has become one of the most popular clustering methods and is available in KNIME.

DBSCAN builds clusters sequentially by selecting a point in p-dimensions and identifying all observations within a distance epsilon, set by the user, of that point. If the number of points is at least a pre-specified minimum (also set by the user), then the original point is labeled a "core point". All such points are identified and are used as the basis for forming clusters. While the process is somewhat involved, there is an excellent video[8] available that describes it graphically.

As with many clustering techniques, no strict rules can be applied to selecting epsilon or the minimum number of points, so some experimentation is usually needed. Published guidance on selecting the minimum number of points varies. Some guidance suggests that the minimum number should be three or more, while other authors suggest that the minimum should be two times the number of dimensions in the data set.

The value of epsilon has a significant effect on the clustering solution. Setting epsilon too small will result in the algorithm identifying a large proportion of the observations as outliers. Setting epsilon too large will result in all observations being in a small number of clusters or even a single cluster. A grid search of DBSCAN results with varying minimum sizes and epsilon values is one approach to determining epsilon, but there is no clear criterion for the optimum result. For example, the silhouette measure discussed in the context of k-means cannot be used with DBSCAN because that metric assumes spherical clusters.

Figure 13.17 shows a pattern of observations that cannot be clustered using hierarchical or k-means. However, experiments have shown that DBSCAN can accurately identify the five clusters.

While DBSCAN has been successfully used in a variety of applications, there are both advantages and disadvantages of the technique.

Advantages of DBASCN include:

- It can discover arbitrarily shaped clusters, including clusters surrounded by different clusters.
- It is robust to outliers and can identify and report noisy observations.
- Only two parameters must be set: epsilon and the minimum number of points.

---

[8] Clustering with DBSCAN, Clearly Explained!!!, url: https://www.youtube.com/watch?v=RDZUdRSDOok&ab_channel=StatQuestwithJoshStarmer

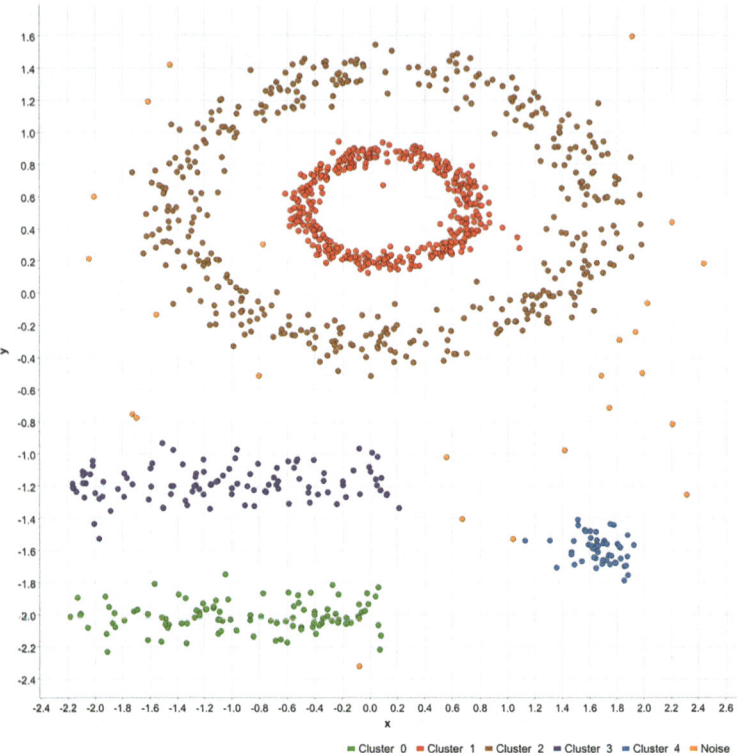

Fig. 13.17 A complex pattern of observations.

Disadvantages include:

- The clustering results are very sensitive to the settings of epsilon, and the choice of epsilon is not deterministic.
- The algorithm does not work well with clusters with varying densities since the parameters will not be optimal for all clusters.
- The algorithm can be slow with large numbers of observations and features.

13.6   Fuzzy Cluster Analysis

With fuzzy clustering (aka soft clustering) each observation has a probability of belonging to each cluster, with values from 0.0 to 1.0 (Bezdek, 1981; Dunn, 1973). With k-means (or hard clustering), each observation is assigned to a single cluster. Fuzzy clustering is used in many of the same application areas as other clustering methods, including medical diagnostics, astronomy, market segmentation, credit

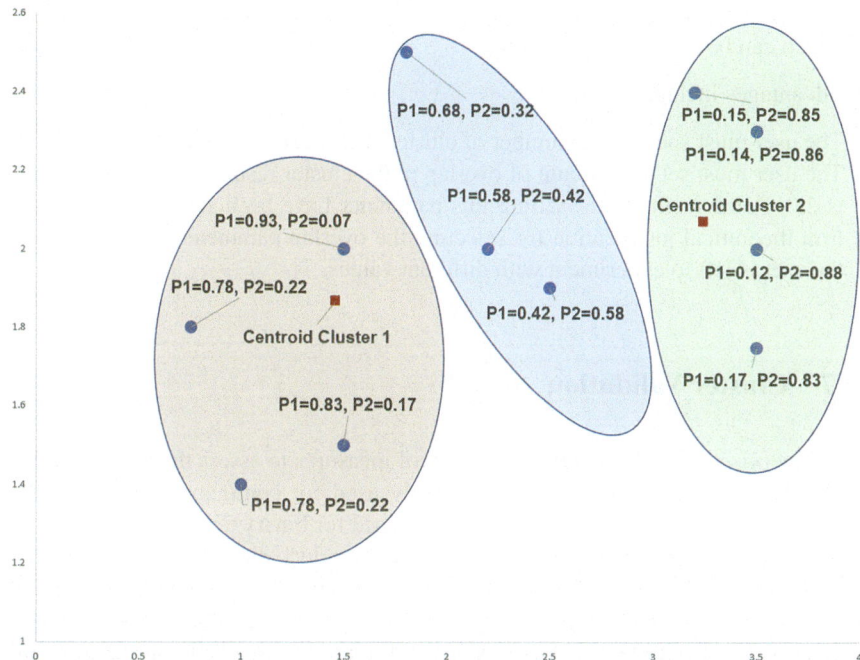

Fig. 13.18 Illustration of fuzzy clustering

scoring, estimation, environmental science, and others. KNIME has a node Fuzzy c-Means for fuzzy cluster analysis.

Figure 13.18 illustrates the fuzzy clustering idea. Eleven observations with two dimensions are plotted in the figure. Two clusters were apparent in the data, with centroids shown for each. The probability of each point belonging to each cluster is indicated. Notice that three points in the center-shaded area have moderate probabilities of belonging to Cluster 1 and Cluster 2. However, ordinary k-means would likely assign the points labeled P1 = 0.68, P2 = 0.32 and P1 = 0.58, P2 = 0.42 to cluster 1 and the point labeled P1 = 0.42, P2 = 0.58 to cluster 2.

Fuzzy c-means requires setting two parameters c, the number of clusters, and m, the "fuzzifier." Higher values of the m result in forcing more clusters to share objects. The minimum value of m is 1.0, which makes fuzzy c-means default to k-means. The typical value of m is 2.0.

As with other clustering techniques, the technique has both advantages and disadvantages.

Advantages of fuzzy clustering include:

- It is robust to outliers.
- Unlike *k*-means, observations are not forced to be in a single cluster, so overlapping clusters can be identified.

- Fuzzy clustering can capture the uncertainty associated with cluster membership which can be an important finding.

Disadvantages include:

- The user must specify the number of clusters before running the model.
- The user must set the amount of overlap in the cluster results. The default value is 2.0. Some methods for setting this parameter have been suggested. Still, no firm theoretical justification for selecting the overlap parameter is available, so the user is left to experiment with different values.

13.7   Cluster Validation

With supervised models, there are a variety of measures to assess the quality of the results. Root mean square error is frequently used for continuous targets, while measures such as accuracy and sensitivity are used for binary targets. An issue with cluster analysis is that clustering algorithms will produce clusters even if there are none in the data. A cluster solution is valid to the extent that the results are not an artifact of the algorithm used and not due to chance. Several approaches to cluster validation have been identified. An extensive literature is available on the topic of validating cluster results. A comprehensive review was published in 2022 (Ullmann et al., 2022).

Internal Cluster Validation
Internal cluster validation relies on the information in the data set being clustered. Hierarchical clustering provides a useful dendrogram to aid cluster validation by explicitly showing which cases are grouped. Hierarchical clustering and k-means can use the Silhouette coefficient as an internal quantitative cluster validity measure. Another internal measure is the within sum of squares which leads to the "elbow plot." However, identifying the elbow in the graph is not always clear, and subjective judgment is required.

Unlike hierarchical clustering, k-means clustering depends upon an initial centroid to begin the clustering process. The initial centroid can be set based on the first k cases or using random initialization governed by a seed. Different seeds can lead to varying numbers of clusters, so if the analysis is run using only a single seed, there is little confidence that a different seed would lead to very different results. An approach to dealing with the dependence on the initialization is to run k-means repeatedly with different seeds and to select the most stable outcome.

External Cluster Validation
In some situations, the analyst has a priori expectations regarding the number and composition of clusters. The degree to which the clustering results conform to the expected structure can be a useful validation criterion. In most cases, however, the number of clusters is not known in advance, so validation can be assessed using variables not part of the set included in the cluster analysis. For example, in a cluster

analysis of voters based on political attitudes, cluster membership could be compared with demographics, such as age, income, education, and political party preference. The degree to which differences are found in these external variables lends credence to the validity of the clusters.

13.8 Summary

Cluster analysis is an unsupervised learning technique to group similar observations into clusters. Key steps include defining the clustering objectives, selecting appropriate variables, checking for clustering tendency, choosing a proximity measure like Euclidean or cosine distance, and selecting a cluster algorithm. Hierarchical clustering builds a tree by iteratively merging clusters using single, complete, or average linkage methods. It does not require specifying the number of clusters a priori.

K-means is a partitioning technique that requires setting the number of clusters k before the analysis is run. It forms spherical clusters by minimizing within-cluster variance. Density-based clustering can find arbitrarily shaped clusters by identifying dense regions. Fuzzy clustering assigns probabilistic membership to each cluster instead of hard cluster assignments. The amount of fuzziness is controlled by a parameter.

In summary, cluster analysis encompasses a variety of techniques to find groups in data, but requires care in algorithm selection, parameter settings, and validation to ensure meaningful clusters are found.

References

Aggarwal, C. C. (2015). *Data mining*. Springer.
Aggarwal, C. C., Hinneburg, A., & Keim, D. A. (2001). On the surprising behavior of distance metrics in high dimensional space. In J. Van den Bussche & V. Vianu (Eds.), *Database theory – ICDT 2001* (Lecture notes in computer science) (Vol. 1973). Springer. https://doi.org/10.1007/3-540-4
Bezdek, J. C. (1981). *Pattern recognition with fuzzy objective function algorithm*. Springer. 4503-X_27. Accessed 29 July 2023.
Dunn, J. C. (1973). A fuzzy relative of the ISODATA process and its use in detecting compact well-separated clusters. *Journal of Cybernetics, 3*(3), 32–57.
Ester, M. H., Kreigel, P., Sander, J., & Xu, X. (1996) *A density-based algorithm for discovering clusters in large spatial databases with noise*. In KDD'96: Proceedings of the second international conference on knowledge discovery and data mining (pp. 226–231). https://dl.acm.org/doi/10.5555/3001460.3001507. Accessed 29 July 2023.
Everitt, S., & Landau, B. S. (2011). *Data mining concepts and technique*. Wiley.
Han, J., Kamber, M., & Pei, J. (2012). *Data mining concepts and techniques*. Morgan Kaufmann Publishers.
Harmouch, M. (2021). *17 Types of similarity and dissimilarity measures used in data science*. https://towardsdatascience.com/17-types-of-similarity-and-dissimilarity-measures-used-in-data-science-3eb914d2681. Accessed 29 July 2023.

Kapri, B. (2019). *Clustering illusion: Clusters lie in the eyes of the beholder!* https://www.absolut-data.com/wp-content/uploads/2019/09/Brainwave-Data-Science-Digest-Sept19.pdf. Accessed 29 July 2023.

Kaufman, L., & Rousseeuw, P. J. (2005). *Finding groups in data.* Wiley.

Lawson, R. G., & Jurs, P. C. (1990). New index for clustering tendency and its application to chemical problems. *Journal of Chemical Information and Computer Sciences, 30*(1), 36–41.

Ullmann, T., Henning, C., & Boulesteix, A. (2022). *Validation of cluster analysis results on validation data: A systematic framework.* https://wires.onlinelibrary.wiley.com/doi/10.1002/widm.1444. Accessed 29 July 2023.

Chapter 14
Communication and Deployment

Creating a predictive model is not the end of an analytics project. Three elements of the "endgame" of analytics are:

- A final written report.
- A presentation based on the final report.
- Deployment of the model.

The level of detail required in the remaining steps depends upon the uses envisioned for the model. The model might be one that only the developer plans to use or perhaps by members of a small team. If the model is to be incorporated into a production process within an organization or made available to outsiders such as potential and current customers, suppliers, or the public, communication and deployment become more involved and important.

14.1 Writing and Presenting the Final Report

Communicating results via written reports and presentations are important aspects of the end-product of the predictive analytics process since insights and conclusions from the project may lead to improvements in an organization, call for additional work, or include plans for deploying a model into production. One of the leading reasons cited in the KDnuggets poll for not deploying an analytic model was found to be: "Decision makers unwilling to approve the change to existing operations" (Siegel, 2022). This unwillingness could be due to poor communication of the benefits of the model, to a lack of understanding of the business situation on the part of the analysts, or a bit of both. Careful upfront understanding of the business problem prior to initiating an analytics project can help ensure that a predictive model will meet cost/benefit requirements.

© The Author(s), under exclusive license to Springer Nature Switzerland AG 2023
F. Acito, *Predictive Analytics with KNIME*,
https://doi.org/10.1007/978-3-031-45630-5_14

There is no surprise about the basic elements of a final modeling report and presentation which should include the following:

- Statement of the original business problem and discussion of any deviations or revisions made to the problem statement as the project evolved.
- The steps followed in the analysis project.
- A summary of the models and findings.
- An indication of whether the models should be deployed and, if so, a plan for deployment.
- Conclusions and recommendations for further work based on the findings.

While these core elements should be present in any report or presentation of a predictive model, the decision of how much and what type of detail to include depends on the audience and the purpose of the project (EMC Educational Services, 2015). Presentations to high-level executives should focus on applications and implications for the business. When presenting to an audience with technical backgrounds, more details on the methodology and analysis can be included.

The goals of communication may include efforts to explain, convey information, persuade, and serve as a call to action. If the model is to be used only by the developer(s), then the final report should include a description of what was done and why, as well as details of the steps taken in the modeling process. The documentation should have enough detail so that if one were to go back weeks or months after creating the model, it can be fully understood so that the model could be modified or passed on to other analysts.

Content for Technical Versus Non-technical Audiences
The main contents of presentations and reports are shown in Table 14.1 for both technical and non-technical audiences.

Style and Order of Presentation Contents
There has been a great deal of discussion about the importance of storytelling in presentations (Dykes, 2020; Knaflic, 2015; EMC, 2015), and (Davenport, 2015). These authors present a convincing case for presenting analytics results in the form of a story. Dykes' book suggests four steps in crafting a data story:

- The setting which provides background on the current situation that culminates on the "hook" – information that reveals a problem or opportunity).
- Insights into the problem or opportunity and supporting details.
- The "Aha Moment" where the major findings are revealed.
- Options, recommended actions, and next steps.

This story structure can be effective with the right audience, but in many business situations the analysts will be faced with executives that are impatient and pressed for time. Many presenters can cite a time when they were cut-off mid-presentation and asked, "What's the point of this?," or, worse, having key executives simply leave the room. When faced with the prospect of an audience like this, the "inverted pyramid" structure is a good alternative to the structured story. The inverted pyramid begins with the "bottom line" and proceeds as follows:

Table 14.1 Content recommendations for technical and non-technical audiences

Content area	Non-technical audience	Technical audience
Project Goals	List the major objectives of the project.	List the major objectives of the project.
Description of the methodology and model	Describe methodology at a high level without technical details. Provide a non-technical overview of the modeling technique. Discuss model accuracy and validity in non-technical terms.	Provide details on modeling techniques and technology and the reason for selecting the particular algorithms. Present the main logic of the model, including data preparation steps, the type of model, list of variables, model type, list of variables, key assumptions, and validation steps, and predictive accuracy.
Discussion of key results	List key findings with bullet points. Avoid text-heavy slides and numeric tables. Instead, use uncluttered, high quality presentation visuals and non-technical graphics.	Show a detailed list of findings. Use analyst-oriented charts and graphs, such as ROC curves, histograms, and density plots.
Recommendations	If deployment is recommended, focus on business impact, including risks and ROI. Otherwise, present recommendations for future efforts.	If deployment is recommended, discuss implications for deploying in a production environment. Otherwise, discuss reasons for non-deployment including model and data limitations. Summarize key learnings for future modeling efforts.

- Most important information, recommendations, and results.
- Supporting information.
- Other details.

The inverted pyramid structure does not mean that engaging anecdotes or examples be omitted or that the presentation to executives should be dry facts and figures. Data visualizations and images can help convey the intended message as with other audiences.

14.2 Data Visualization

Data visualization is an important part of both written and oral reports. There are two main uses for graphical display of data: exploration and presentation. KNIME provides many options for data visualization to support data cleaning, exploration, and understanding. KNIME graphics enable users to visually explore data to discover new insights, but those graphics are not always presentation ready. Instead, the data outputs from KNIME analytics can be exported to Excel, Power BI, Tableau, and other software that can produce presentation-ready and interactive graphics and dashboards. There are many data visualization software packages available, some open-source, and some quite expensive.

While nearly everyone learned about pie charts and bar charts in elementary school, the power of computer processing and high-resolution monitors and color printers have created new opportunities for analysis and communications. These new modes of display can leverage the power, flexibility, and immediacy of the human perceptual process. Unfortunately, the technology has also led to the ability to create graphs and charts that are more likely to confuse or mislead than communicate by including chart junk, e.g., unnecessary background images, background colors, 3-d effects, bold labels, redundant labels, borders, and shadows (Raja, 2022). As a rule, anything that does not contribute in an essential way to the meaning of a graph is a distraction that harms communication and should be removed.

There are several tasks involved in creating a visual display. The first task is data selection. This refers to the art of selecting which data to include in the visual representation. This is clearly an important step because data left out can result in misleading conclusions, but it is also important because all data and details cannot be included.

The second task is encoding the data into graphical form which involves mapping data into visual cues. The objective of mapping is to create an accurate visual representation of the data so that interpretations are faithful to the actual data. The importance of this step is made evident by the many examples of purposefully or accidentally created graphics that confuse or even mislead viewers.

The third task is constructing the graphic image in the available space on the page or the screen in a way that accurately and effectively communicates to the audience. There are four basic purposes for creating data visualization: making comparisons, showing composition, demonstrating relationships, and showing distributions.

The fourth task is incorporating interactivity and motion into the visualization. While this is not always feasible, interactivity can increase the effectiveness of what is communicated as well as enhance audience engagement. Software has been developed and made available that can make visualizations dynamic so that the user or presenter can manipulate what is seen and dig into the data at will to refine or focus understanding.

Encoding the Data into Graphical Form
The design and structure of a graph has a strong effect on what is perceived and what is attended to. The French cartographer Jacques Bertin developed the conceptual foundations for graphical design in his classic work Sémiologie Graphique published 1967. This groundbreaking work, later translated into English (Bertin, 1983), organized the visual elements of graphical display according to the type of information that is communicated. Bertin identified the main categories of elements of visualization that can be used to convey quantities:

Table 14.2 Examples of chart types

For comparison		For composition	
Line charts	Radar charts	Stacked bar charts	Sankey diagrams
Bar charts	Heat maps	Tree maps	Pie and donut charts
Stacked bar charts	Population pyramids	Waterfall charts	Sunburst diagrams
Dot charts	Quadrant maps	Stacked area charts	Venn diagrams
Slope charts	Candlestick charts	Marimeko charts	Dendrograms
		Funnel charts	Pipeline charts
For distributions		For relationships	
Density plots	Pareto charts	Scatterplots	Network charts
Histograms	Dot matrix charts	Scatterplot matrices	Trellis charts
Box plots		Bubble charts	Dot maps
Violin plots		Chord diagrams	Choropleth maps
Stem and leaf plots			

- Shape
- Curvature
- Size
- Orientation
- Color
- Texture
- Line length and width
- Color intensity
- Position
- Blur

These elements can be used to construct an enormous variety of charts. Selected examples are shown in Table 14.2. Examples of charts and software to create them, including open-source tools, are widely available.

Constructing the Graphical Image
While there are some principles that have been offered by graphics experts such as Stephen Few and Edward Tufte, creating a visualization is best considered a craft rather than a science. Visualization has roots in statistics, graphic design, cartography, cognitive science and computer science and this multi-disciplinary mix makes it a difficult craft to master. There is not a generally agreed-upon set of principles, yet there are valuable resources available, e.g., Kirk (2012).

14.3 Deploying Predictive Models

Deploying predictive models, while part of the analytics process discussed in Chap. 1, can be a separate and complex process. It is understandable that the analyst or data sciences team would like to see the model used after spending the time and effort to build it. The scope of deployment can be:

- For use by the developer or development team.
- Incorporated into a production process within the organization's systems.
- Made available to current and potential customers, suppliers, investors, or others.

In addition, the data pipeline for a predictive model can be in batch mode as needed or scheduled, or available for real time processing. As implementation ranges from individual, in-company use to real-time processing for external users, the complexity, investment, and need for buy-in all become more important. As the deployment scope moves from batch use by the developer or development team toward real-time use outside of the organization, complexity and required investments increase in:

- Data privacy and security for both users and organizations.
- Robustness to variations in input data.
- Usability of the interface.
- The need for periodic maintenance.
- Performance monitoring since the assumptions used to create the model and environment may change over time.

Not all projects will lead to deployment. In fact, there is substantial evidence that most models are not successfully deployed. A KDnuggets poll revealed that many data scientists reported that fewer than 20% of models generated were deployed (Siegel, 2022). Dean Abbott noted the reasons that many predictive models are never deployed include technical issues, political roadblocks, or government regulations (Abbott, 2014). This does not necessarily mean that the project has no value because the insights can still lead to changes in strategy.

Abbott identified several approaches to deployment. The simplest, and most common, is to deploy a model using the predictive modeling software itself. This can be done in-house with KNIME and does not require any further investment in software or services. The users need to have KNIME installed on their machines and access to the relevant workflows. The models should not be distributed outside the user group, however.

Deployment to constituencies outside of the organization is usually done via cloud services. Some software solutions offer add-ons to deploy models. KNIME, IBM SPSS Modeler, Rapid Miner, Amazon ML, SAS Enterprise Model, and other services can deploy models without the need for developing custom software. While these approaches are convenient and frequently offer features such as automatic alerts for problems, these services can be expensive.

Other deployment approaches typically require translation of the model into languages such as C#, Java, or Python. PMML (Predictive Model Markup Language) is a standard that allows communication of models from analytics software (such as KNIME, Rapid Miner, SAS, SPSS, etc.) by translating a predictive model into a standard format which can then be read by other programs for deployment. The PMML files contain the data dictionary, the data transformations used in the predictive model, and information on the specific model (e.g., decision trees, regression), and outputs with the predictions.

The competencies required for moving a model into production are different from the skillset and experience of most analysts and thus a handoff to IT teams is usually necessary. Since the emphasis in this book is on low- or no-code analytics, only in-software deployment will be demonstrated using KNIME.

Deploying Predictive Models for Use by the Developer or Team
There are two simple ways of deploying a model for individual or team use. One is to deploy the model using the software tool used to create the model. The other is to translate the model into another tool or program. Examples of each approach will be presented here.

Deployment Using KNIME
KNIME offers Integrated Deployment nodes to capture the workflows needed to run in a production environment. This example will demonstrate deploying a workflow within the KNIME software itself.

The example from Chap. 7 on predicting employee retention will be used in this demonstration. Figure 14.1 shows a workflow which captures two elements of a logistic regression model to predict employee retention: (a) data preparation which involves taking the logarithm of Years at the Company, and (b) the logistic model predictor using the input data. The two captured areas are combined and then saved as workflow.

In general, new data submitted to a saved model must be prepared and processed in the same way that was done for model building. This includes how missing values were handled, variable transformations, normalization, and other data preparation steps. The saved workflow can then be used to score new data. While this example only has a single Math node, it is important that all data preparation steps are captured in model deployment workflow so that model performance is preserved when new data is applied. Table 14.3 has the descriptions of each node in Fig. 14.2.

At this point a new workflow has been automatically created which can use data on employee characteristics to predict retention. The only node to add is the File Reader with the new input data. The variable in the input data file must use the same formats and column names as the original data used to create the workflow in Fig. 14.1. The created workflow is shown in Fig. 14.2. The nodes within the "Saved deployment workflow" box were created by KNIME when the workflow was written to WorkflowForFig.14.2. The Test input data box contains a single node that is used to read new data. In this case a File Reader node was inserted to read data from

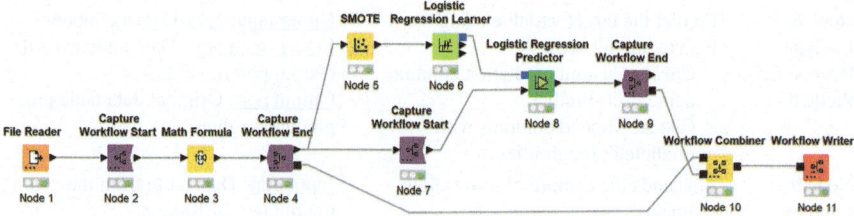

Fig. 14.1 Capturing the employment retention model for deployment

Table 14.3 Node descriptions for the Fig. 14.1 workflow

Nodes	Descriptions	Input and output ports
Node 1 File Reader	Read Employee_turnover.csv.	Output port: KNIME data table with employee turnover with 14,999 rows and 9 columns.
Node 2 Capture Workflow Start	This starts the capture of a workflow segment. The entire workflow within the scope of the Capture Workflow Start and Capture Workflow End nodes are made available.	Input port: Data table from the Output port of Node 1. Output port: Captured workflow inputs plus KNIME data table with employee turnover with 14,999 rows and 9 columns.
Node 3 Math Formula	This node can evaluate a mathematical expression based on the row values of a variable using a large assortment of available functions. Math Expression: ln($YearsAtCompany$). Append Column: lnYearsAtCompany.	Input port: Data table from the Output port of Node 2. Output port: Data table with ln(YearsAtCompany).
Node 4 Capture Workflow End	This ends the capture of a workflow segment.	Input port: Data table from the Output port of Node 3. Output port: Captured workflow inputs plus Data table with ln(YearsAtCompany)
Node 5 Smote	SMOTE oversamples the input data. Settings: Class column: Left. Check: Oversample minority classes. Check: Enable static seed: 123.	Input port: Data table from the Output port of Node 3. Output port: Oversampled with a balanced number of Left and Did not leave target values.
Node 6 Logistic Regression Learner	Create the logistic model. Settings: Target column: Left. Select solver: Iteratively reweighted least squares. Include: All predictor variables except YearsAtCompany.	Input port: Data table from the Output port of Node 5. Upper output port: Logistic regression model. Middle Output port: Coefficients and statistics. Lower output port: Model and learning properties.
Node 7 Capture Workflow Start	This starts the capture of a workflow segment. The entire workflow within the scope of the Capture Workflow Start and Capture Workflow End nodes are made available.	Input port: Data table from the Output port of Node 4. Output port: Captured workflow inputs plus KNIME data table with employee turnover with 14,999 rows and 9 columns.
Node 8 Logistic Regression Predictor	Predict the target variable using the logistic model settings: Check: Custom prediction column name: Left Predict. Check: Append columns with predicted probabilities.	Upper input port: Logistic model Lower input port: Data table from the Output port of Node 7. Output port: Original data table plus predicted values.
Node 9 Capture Workflow End	This ends the capture of a workflow segment.	Input port: Data table from the Output port of Node 8. Output port: Captured workflow inputs plus original data table plus predicted values.

(continued)

Table 14.3 (continued)

Nodes	Descriptions	Input and output ports
Node 10 Workflow Combiner	Combines two workflows into a single workflow.	<u>Upper input port:</u> Captured workflow from Node 4. <u>Lower input port:</u> Captured workflow from Node 9. <u>Output port:</u> Combined workflows.
Node 11 Workflow Writer	Write the workflow to the current workflow area. Output location: Write to: Relative to Current workflow Folder: ../deploylogisticmodel Write options: Create missing folders Workflow: If exists: Select Overwrite Check: Use custom workflow name. Custom workflow name: turnover.	<u>Input port:</u> Workflow from the Output port of Node 10.

Fig. 14.2 Deployed workflow to predict employee retention

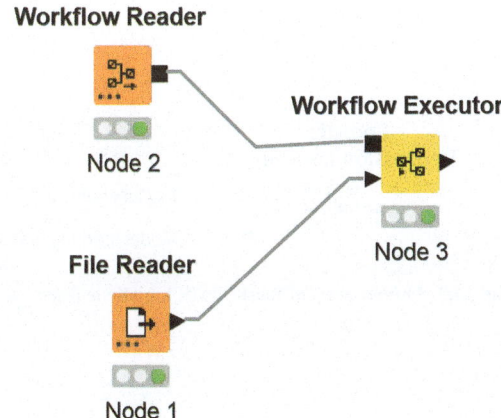

a .csv file. The new data for this demonstration consisted of five rows of employee characteristics which are shown in Table 14.4.

The predicted values for the five test cases are listed in Table 14.5. This approach would work for any number of new cases. Any analyst or manager wishing to use this model would need to have KNIME installed on their machine plus access to the file WorkflowForFig.14.2.

Deployment Using Excel
If the model uses a relatively straightforward formula to make predictions, the model can be deployed in Excel. The logistic equation model from the workflow in Fig. 14.2 is available from the Logistic Regression Learner node. This node has three output ports as shown in Fig. 14.3.

Table 14.4 Test input data to predict employee retention

Row	Satisfaction	Last evaluation	Projects	Average hours	Years at company	Promotion	Department	Salary
0	0.7	0.88	2	157	3	No	Sales	High
1	0.2	0.5	5	159	5	Yes	IT	Low
2	0.2	0.7	4	122	5	Yes	Sales	High
3	0.2	0.5	3	144	5	Yes	Technical	Low
4	0.2	0.5	2	159	5	No	HR	Medium

Table 14.5 Test input data to predict employee retention

Row	Probability of leaving	Probability of not leaving	Prediction
0	0.14	0.86	Will not leave
1	0.49	0.51	Will not leave
2	0.22	0.78	Will not leave
3	0.78	0.22	Will leave
4	0.94	0.06	Will leave

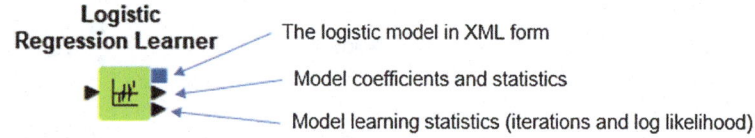

Fig. 14.3 Output ports of the KNIME logistic learner

For this example, the model coefficients were captured from the middle port (shown in Table 14.6) and copied into an Excel workbook. An Excel worksheet was created that used the coefficients to estimate the probability of leaving as a function of the values of the input features. The worksheet, shown in Fig. 14.4 was set to check values in each column to ensure that they were in the appropriate range. For example, Satisfaction ranged from 0.0 to 1.0, so any values outside this range would not be accepted.

As inputs are entered into Excel worksheet, the Probability of leaving is calculated, and the Prediction is made. The worksheet can handle as many rows of input data as needed.

Table 14.6 Logistic
regression coefficients from
the workflow in Fig. 14.2

Variable	Coefficients
Satisfaction	−4.761
LastEvaluation	1.023
Projects	−0.527
AverageHours	0.005
Promotion=Yes	−1.732
Department=ResAndDevel	−0.397
Department=accounting	0.185
Department=hr	0.422
Department=management	−0.459
Department=marketing	0.127
Department=product_mgt	0.076
Department=sales	0.135
Department=support	0.225
Department=technical	0.301
Salary=Low	1.927
Salary=Medium	1.413
LN_YearsAtCompany	2.556
Constant	−2.042

Satisfaction (0 to 1.0)	Last evaluation score (0 to 1.0)	Projects (1 to 10)	Average Hours (50 to 350)	Promotion (Yes or No)	Salary Low, Medium, or High	Years at the company (1 to 10)	Department R&D / Accounting / HR / Management / Marketing / Product management / Sales / Support / Technical / IT	Probability of leaving	Prediction
0.80	1.00	2	90	No	High	3	IT	0.21	Will not leave
0.55	0.80	6	122	Yes	Medium	9	Sales	0.98	Will leave
0.60	1.00	2	66	No	High	2	IT	0.18	Will not leave
0.76	0.70	6	300	No	Low	3	Management	0.26	Will not leave
0.90	0.75	6	321	Yes	Medium	5	Accounting	0.84	Will leave
0.95	0.80	3	56	Yes	Medium	3	HR	0.65	Will leave
0.70	0.65	3	200	No	High	4	Management	0.29	Will not leave

Fig. 14.4 Segment of Excel worksheet for predicting retention

14.4 Summary

The final steps of an analytics project are communicating the results via a report, presentation to an audience, and deploying the model. Reports and presentations should cover the business goals, methodology, key findings, recommendations, and next steps at the appropriate level of detail for the audience.

Presentation tips include leading with the most important information, using graphics and visuals, and structuring content in a story format or inverted pyramid style depending on the audience. Data visualization encodes data into visual elements like position, length, color and shape to explore data and create effective graphics for communication.

Deploying models ranges from individual use to real-time processing at scale, with increasing complexity. Capturing workflows and using model export formats simplifies deployment. KNIME enables model deployment through captured workflows and nodes like the Integrated Deployer. Models can also be deployed using other platforms like Excel.

Considerations for full-scale deployment include data privacy, robustness, usability, maintenance and monitoring. Many models are never deployed due to technical or business issues.

In summary, communicating results convincingly and deploying models successfully require focus on audience needs, clear visuals, and managing model integration complexity.

References

Abbott, D. (2014). *Applied predictive analytics*. Wiley.
Bertin, J. (trans. W. J. Berg) (1983). *Semiology graphics: diagrams, networks, maps*. The University of Wisconsin Press,
Davenport, T. (2015). *Why data storytelling is so important – And why we are so bad at it*. https://www.linkedin.com/pulse/why-data-storytelling-so-importantand-were-bad-tom-davenport/. Accessed 29 July 2023
Dykes, B. (2020). *Effective data storytelling: How to drive change with data, narrative and visuals*. Wiley.
EMC Education Services. (2015). The endgame, or putting it all together. In EMC Education Services (ed) *Data science & big data analytics: Discovering, analyzing, visualizing and presenting data* (pp. 359–395). Wiley.
Kirk, A. (2012). *Data visualization: A successful design process*. Packt Publishing.
Knaflic, C. N. (2015). *Storytelling with data*. Wiley.
Raja, A. A. (2022). *Data visualization best practices*. https://www.linkedin.com/pulse/data-visualization-best-practices-ammar-a-raja/. Accessed 29 July 2023
Siegel, E. (2022). *Models are rarely deployed: An industry-wide failure in machine learning leadership*. KDnuggets. https://www.kdnuggets.com/2022/01/models-rarely-deployed-industrywide-failure-machine-learning-leadership.html. Accessed 29 July 2023

Index

© The Editor(s) (if applicable) and The Author(s), under exclusive license to   311
Springer Nature Switzerland AG 2023
F. Acito, *Predictive Analytics with KNIME*,
https://doi.org/10.1007/978-3-031-45630-5

SPRINGER NATURE

GPSR Compliance

The European Union's (EU) General Product Safety Regulation (GPSR) is a set of rules that requires consumer products to be safe and our obligations to ensure this.

If you have any concerns about our products, you can contact us on ProductSafety@springernature.com

In case Publisher is established outside the EU, the EU authorized representative is:

Springer Nature Customer Service Center GmbH
Europaplatz 3
69115 Heidelberg, Germany

The manufacturer's authorised representative in the EU is Springer Nature Customer Service Centre GmbH, Europaplatz 3, 69115 Heidelberg, Germany. If you have any concerns regarding our products, please contact ProductSafety@springernature.com

Printed and bound by CPI Group (UK) Ltd, Croydon, CR0 4YY

23/04/2026

02095594-0010